Leckie✕Leckie
Scotland's leading educational publishers

Success guides

HIGHER
PHYSICS

✕ JOHN IRVINE ✕

Contents

Mechanics and Properties of Matter

Electricity and Electronics

Contents

Radiation and Matter

Scalars and Vectors

Types of Quantity

In Physics there are many measurable quantities, such as; time, length, force, weight, temperature, speed, acceleration, mass, charge, velocity and energy. Every quantity is classified as being either a **scalar** quantity or a **vector** quantity. When the quantity **needs a direction**, it is called a **vector** quantity. When a direction is **not** needed, the quantity is a **scalar** quantity. Another way of saying this is that if a quantity has only size (or magnitude), then it is a scalar quantity; if it has magnitude and direction, then it is a vector quantity.

Top Tip

Make your own list of scalars and vectors for the whole course.

Examples

Scalar Quantities
(no direction needed)

    speed
charge energy
temperature
        mass
    time

Vector Quantities
(direction needed)

    weight
        velocity
force
    acceleration

Weight is a force towards the centre of the earth. Velocity is a speed plus the direction of travel. The full effect of a force cannot be worked out without knowing its direction as well as its value. An example of the direction of acceleration is that the acceleration due to gravity is vertically downwards.

Giving directions

One way of quoting a direction is to use the points of the compass, i.e.

North
West ←——→ East
South

Example

The direction of the red arrow could be stated either as 35° north of west OR 55° west of north.

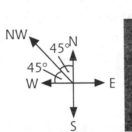

It is important to note that the direction 'north-west' is correct only for the direction exactly halfway between N and W.

Three-figure bearings

To give the direction of a vector as a three-figure bearing, firstly imagine that you are at the start of the vector **facing due North**, then **turn clockwise** until you face along the vector's direction. The **angle you turn through** is the required bearing.

This means, for example, that:
- due north has a bearing of 000°
- due east has a bearing of 090°
- due south has a bearing of 180° and
- due west has a bearing of 270°.

N = 000°
W = 270° ←——→ E = 090°
S = 180°

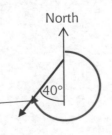

Example
The bearing of this vector is 180 + 40 = 220°.

Distance and displacement

Distance, *d*, is a scalar quantity. It is measured in metres (m).

Example of distance

Mother says, 'Take the dog for a 3.0-km walk.' The dog has to be walked a distance of 3.0 km in order to get exercise. Direction is unimportant; the dog could be walked round and round in circles until the total distance of 3.0 km was achieved.

Displacement, *s*, is a vector quantity. It is a distance plus a direction.

Example of displacement

 'The treasure is buried 50 m due west of the old oak tree.' Here, direction is essential. To leave out the 'due west' information would make it very difficult to find the treasure.

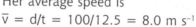

50 m

Speed and velocity

Average speed, $\bar{v}$, is defined as the distance travelled each second.

Speed is a scalar quantity. It is usually measured in metres per second (m s^{-1}).

Example of speed

The police stop a motorist travelling at 50 m.p.h. (the speed) along a high street in a town. The police are only interested in how fast the car was going; direction is irrelevant.

Formula for speed

$$\text{average speed} = \frac{\text{distance}}{\text{time}}$$

$$\bar{v} = \frac{d}{t}$$

Example

A cyclist travels 100 m in 12.5 seconds. Her average speed is
$\bar{v} = d/t = 100/12.5 = 8.0$ m s^{-1}

Top Tip

Practise using the speed and velocity formulas in questions which ask you to find the time or speed or velocity.

Velocity is a vector quantity. It is a speed plus a direction.

Example of velocity

An aircraft needs to reach an airport on schedule. Air traffic control instructs the pilot to fly at 150 m s^{-1} bearing 135° (the velocity). The plane must fly both at the correct speed and in the correct direction in order to meet the schedule.

Formula for velocity

$$\text{average velocity} = \frac{\text{displacement}}{\text{time}}$$

$$v = \frac{s}{t}$$

Example

A ship travels 90 km due south in 12 hours. Its average velocity is $v = d/t = 90/12$
$= 7.5$ km h^{-1} bearing 180°.

Quick Test 1

1. What is the difference between scalar quantities and vector quantities?
2. Classify 'velocity', 'distance', 'displacement' and 'speed' as scalars or vectors.
3. A plane is flying on a course 25° west of north. What is this as a three-figure bearing?
4. A car travels 18 km in 30 minutes. Calculate its average speed in metres per second.
5. An orienteer takes 2 hours to reach a checkpoint which is 5 km away from the starting point. The checkpoint is 18° south of east from the starting point. Calculate the average velocity.

Adding together Vector Quantities

Adding vectors along the same straight line

When vectors are pointing in the same direction they are simply added together.

$$\xrightarrow{8\ N} + \xrightarrow{5\ N} = \xrightarrow{13\ N}$$

When vectors are pointing in opposite directions they are subtracted.

$$\xrightarrow{8\ N} + \xleftarrow{5\ N} = \overset{3\ N}{\downarrow}$$

Adding vectors in two dimensions

When vectors are at an angle to each other, normal arithmetic does not apply and the sum (resultant) can be found by scale diagram (or geometry and trigonometry if your maths is good enough!).

To combine vectors by scale diagram

(a) Choose a suitable scale. (The diagram needs to be large enough or it will be inaccurate.)
(b) Choose a suitable point to start drawing the vectors – this is called the origin.
(c) Draw the first vector to scale and pointing in the correct direction.
(d) At the **end** of the first vector draw the second vector to scale and in its correct direction.
(e) Draw a straight line from the origin to the finishing point – this is the resultant.

Example 1: (Adding two displacement vectors)

A cross-country runner jogs 4.0 km due north and then turns and runs due west for a further 3.0 km. What is the runner's final displacement?

(a) A suitable scale could be 1 cm = 1.0 km.
(But note that a scale of 1 cm = 0.5 km would make a bigger diagram which would be more accurate.)

(b) The origin should be near the bottom of the page on the right hand side – this allows space to draw the vectors upwards (north) and to the left (west).

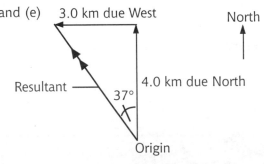

Measurement on the diagram gives the length of the resultant vector to be 5.0 cm which is equivalent to 5.0 km in real life.

A protractor should be used to measure the shown angle. This angle is 37° west of north, which is equivalent to a three-figure bearing of 323°.

Finally, the resultant should be quoted as '5.0 km bearing 323°'.

Top Tip

Make sure you show arrows on all the vector lines in your diagram.

Using geometry and trigonometry

West and north are at right angles to each other. This means the triangle is right-angled and Pythagoras' theorem can be used to find the hypotenuse, i.e.

Resultant = $\sqrt{(3^2 + 4^2)} = \sqrt{(9 + 16)} = \sqrt{25} = 5.0$ km.

Tan x = opposite/adjacent = $\frac{3}{4} \Rightarrow x = \tan^{-1} 0.75 = 37°$.

Example 2

A person walks 5.0 km due south in 2.0 hours.

He then turns and walks a further 10.0 km on a bearing of 070°, taking a further 3.0 hours.

Calculate:

(a) the total distance travelled;

(b) his average walking speed;

(c) his resultant displacement;

(d) his average velocity.

Answers

(a) Total distance = 5.0 + 10.0 = 15.0 km.

(b) Average speed = total distance/time taken = 15.0/5 = 3.0 km per hour.

(c) Scale 1.0 cm = 1.0 km.

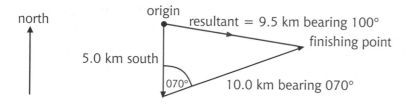

Note that a ruler and protractor are needed to find the size and direction of the resultant. To find the solution using trigonometry, you need to know both the cosine rule and the sine rule. For many students it is easier to use a scale diagram.

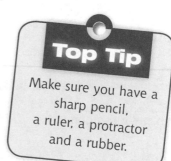

Top Tip

Make sure you have a sharp pencil, a ruler, a protractor and a rubber.

(d) Average velocity = total displacement/time = 9.5/5 = 1.9 km h⁻¹ bearing 100°.

Quick Test 2

1. What is the resultant force acting on the following object?

2. What is the resultant force on the following toy?

3. A model plane flies 150 m due south and then 200 m due east. Calculate its resultant displacement.

4. A velocity of 5.0 m s⁻¹ due west is combined with a velocity of 12.0 m s⁻¹ due north. Calculate the resultant velocity.

5. A force of 16 N bearing 090° is combined with a force of 25 N bearing 050°. Calculate the resultant force.

Hint:

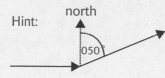

Resolving Vectors into Components

Resolving vectors

Resolving a vector means breaking it down into two vectors. These two vectors are usually chosen to be at right angles to each other. A vector, v, at an angle of $\theta°$ to the horizontal can be resolved into its horizontal component, v_x, and its vertical component, v_y, as follows.

$$\cos \theta = \frac{\text{adjacent}}{\text{hypotenuse}} = \frac{v_x}{v}$$

$\Rightarrow$ | the horizontal component of the vector, $v_x = v \cos \theta$ |

$$\sin \theta = \frac{\text{opposite}}{\text{hypotenuse}} = \frac{v_y}{v}$$

$\Rightarrow$ | the vertical component of the vector, $v_y = v \sin \theta$ |

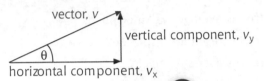

vector, v

vertical component, v_y

θ

horizontal component, v_x

Top Tip

You need to memorise these relationships – they are not listed in the Physics data booklet!

Example

A force of 80 N acts at an angle of 35° up from the horizontal. What are the horizontal and vertical components of this force?

Answer

80 N

vertical component, v_y

35°

horizontal component, v_x

the horizontal component of the vector, $v_x = v \cos \theta$
= 80 cos 35°
= 65.5 N

the vertical component of the vector, $v_y = v \sin \theta$
= 80 sin 35°
= 45.9 N

All types of vector quantities can be resolved like this into their components. It is often useful, for example, to break down the initial **velocity** of a projectile into its horizontal and vertical components.

Example

An athlete throws a javelin with a speed of 24 m s⁻¹ at an angle of 48° to the horizontal.
Calculate the horizontal and vertical components of its initial velocity.

24 m s⁻¹

48°

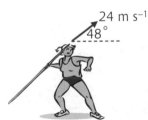

Answer

Horizontal component of the initial velocity = $v \cos \theta$ = 24 × cos 48 = 16.1 m s⁻¹.
Vertical component of the initial velocity = $v \sin \theta$ = 24 × sin 48 = 17.8 m s⁻¹.
i.e.

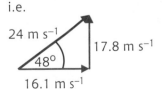

24 m s⁻¹

17.8 m s⁻¹

48°

16.1 m s⁻¹

Top Tip

The section on projectiles (page 16) shows how these values can then be used to calculate maximum height, time of flight and horizontal range.

Object on a slope

Any object on a slope tends to run (or slide) downhill.

This is because of its **component of weight** acting down the slope.

To find this component of weight down the slope, the weight vector must be resolved into two components, one parallel to the slope (the one we want) and one perpendicular to the slope, i.e.

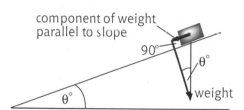

component of weight parallel to slope

weight

From the right-angled triangle:

$$\sin \theta = \frac{\text{opposite}}{\text{hypotenuse}}$$

$$= \frac{\text{component of weight down slope}}{\text{weight } (=mg)}$$

This produces the following relationship,

| Component of weight down slope = m g sin θ |

('g' is the gravitational field strength, [= 9.8 N kg⁻¹ on earth])

N.B. This relationship is **not** listed in the Physics data booklet – you need to memorise it for the exam.

Top Tip

Remember that in many questions you will also need to take into account other forces (like friction).

Example

A cyclist and bicycle have a total mass of 110 kg.

The cyclist pedals uphill on a slope which makes an angle of 8° to the horizontal.

The cyclist exerts a driving force of 260 N. The force of friction is 40 N.

(a) Calculate the component of weight parallel to the slope.

(b) Calculate the unbalanced force.

Answer

(a) Component of weight down slope = m g sin θ = 110 × 9.8 × sin8° = 150 N.

(b) Unbalanced force = driving force – (compt. of weight + friction) = 260 – (150 + 40) = 70 N.

Quick Test 3

1. A gardener exerts a pushing force of 160 N down the handle of a lawnmower. The handle makes an angle of 65° to the horizontal. Calculate the horizontal component of the pushing force.
2. A golf ball leaves the tee with an initial velocity of 58 m s⁻¹ at an angle of 36° to the horizontal. Calculate the horizontal and vertical components of this velocity.
3. A van of mass 2,400 kg is on a slope which makes an angle of 12° to the horizontal. Calculate the component of the van's weight parallel to the slope.
4. A box of mass 42 kg is placed on a 16° slope. The maximum force of friction between the box and the slope is 124 N. Does the box slide down the slope?
5. In the example further up this page, the cyclist now turns and pedals **down** the same slope. Each force has the same size as before. What is the new unbalanced force?

160 N
65°

m g sin ϑ
12°

Acceleration

Acceleration and velocity

An object is accelerating when its **velocity is changing.** The value of its acceleration means how much the velocity changes each second.

The formula used to calculate acceleration, a, is:

$$a = (v - u)/t$$

where v = final velocity (in m s^{-1})
u = initial velocity (in m s^{-1})
t = the time (in s) for the velocity change.

(You can remember that 'u' stands for the **initial** velocity by remembering that 'u' comes before 'v' in the alphabet.)

The units for acceleration are m s^{-2} (metres per second squared). Acceleration is a **vector** quantity.

Top Tip

You must make sure you substitute the values for 'u' and 'v' the correct way round in the formula.

Example

The speed of a train changes from 42 m s^{-1} to 16 m s^{-1} in 2 minutes. Calculate the acceleration of the train.

Answer

$$a = \frac{(v-u)}{t} = \frac{(16-42)}{120} = \frac{-26}{120} = -0.22 \text{ m s}^{-2}$$

Note that the acceleration is negative; this shows that the train is decelerating.

Using Light Gates to measure acceleration

A light gate consists of a lamp and a light sensor.

A card is attached to the moving object.

As the card passes between the lamp and sensor it breaks the beam of light. This produces signals which switch a timer on and off.

A single card with **one** light gate can be used to find the instantaneous speed of the object as it passes the gate. The formula is:

$$\text{speed} = \text{length of card/time}$$

A single card with **two** gates can be used find the acceleration of an object.
Gate 1 is used to find the initial velocity, u.
Gate 2 is used to find the final velocity, v.
The time, t, to travel from gate 1 to gate 2 must also be measured.
The acceleration is then calculated from:

$$a = (v - u)/t$$

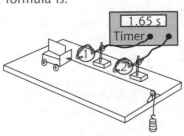

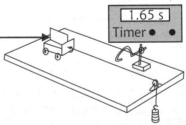

It is also possible to measure the acceleration of an object by using a single light gate and an object with a **double** card attached to it. The attached electronic timer uses the different sections of the card to find the initial velocity, the final velocity and the time for the velocity change.

It can then give a direct readout of the object's acceleration.

The acceleration due to gravity, g, on earth

When the effects of air resistance are ignored, all objects fall with the **same** value of **constant acceleration.** This means the vertical velocity increases by the same amount each second and produces the following time-lapse photograph.

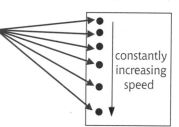

Images of a falling ball taken in equal steps of time

constantly increasing speed

Galileo discovered that different objects took the same time to fall to the ground from Pisa's tower.

A hammer and a feather being dropped together on the Moon. They took the same time to fall.

The acceleration due to gravity can be measured using an object carrying a double card falling through a light gate, i.e.

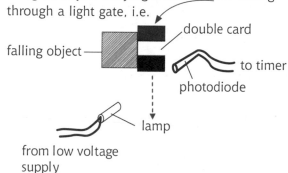

falling object

double card

to timer

photodiode

lamp

from low voltage supply

Example

Using a card with opaque sections each of length 4.0 cm (= 0.04 m).

(a) Time for first break of light beam = 62 m s.
(b) Time for second break of light beam = 37 m s.
(c) Time between breaks = 0.045 s

$\Rightarrow$ initial velocity, u = 0.04/0.062 = 0.645 m s^{-1}
  final velocity, v = 0.04/0.037 = 1.081 m s^{-1}
  acceleration = (v − u)/t = (1.081 − 0.645)/0.045 = 9.7 m s^{-2}.

Top Tip

In Higher Physics, always use a value of 9.8 m s^{-2} for the acceleration due to gravity on the earth.

Quick Test 4

1. 'An object with a high acceleration must be travelling very fast.' Is this statement true or false?
2. 'Two objects can be travelling at different velocities but have the same acceleration.' Is this statement true or false?
3. What does an acceleration of 4.0 m s^{-2} mean?
4. A cyclist's speed increases from 2.6 m s^{-1} to 11.7 m s^{-1} in a time of 6.5 s. Calculate the acceleration of the cyclist.
5. A car is travelling at a velocity of 37.0 m s^{-1}. The driver now applies the brakes. After braking uniformly for a time of 3.4 s, the car's velocity is 21.7 m s^{-1}. Calculate the car's acceleration.
6. A stone is dropped over the edge of a high cliff. How fast is it falling 3.0 s after being released?

Graphs of Motion

Three types of graphs of motion

There are three types of graphs of motion you should know about and understand. These are displacement/time graphs, velocity/time graphs and acceleration/time graphs. One possible experimental arrangement used to produce these graphs is shown below.

The motion sensor sends out pulses of ultrasound. It also detects the echoes. The computer analyses the time delay for the echoes and draw the graphs.

Graphs for constant speed

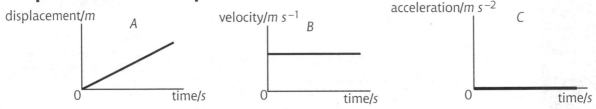

A – a straight diagonal line because a constant speed means equal distances are travelled in equal steps of time.

B – a horizontal line because the velocity is the same at all times.

C – a horizontal line at zero acceleration because the velocity is **not changing.**

Graphs for constant positive acceleration

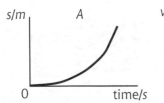 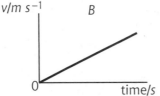 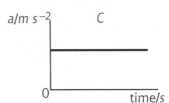

A – a line curving upwards because increasing speed means greater distances are travelled in equal steps of time.

B – a straight diagonal line sloping upwards because constant positive acceleration means velocity increases by equal amounts in equal steps of time.

C – a horizontal line because the acceleration is the same at all times.

Graphs for constant negative acceleration

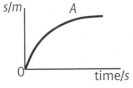 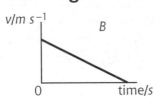 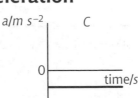

A – a line curving downwards because decreasing speed means smaller distances are travelled in equal steps of time.

B – a straight diagonal line sloping downwards because constant negative acceleration means velocity decreases by equal amounts in equal steps of time.

C – a horizontal line because the acceleration is the same at all times. The line is below the time axis because the value of acceleration is negative.

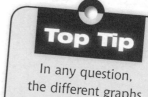

Top Tip

In any question, the different graphs cannot be the same shape (e.g. diagonal straight lines).

Important information

Important information that can be calculated from graphs of motion.

- The **displacement**, s, is the **area** under a velocity/time graph.
- The **acceleration**, a, is the **gradient** of a velocity/time graph.

Example

The velocity/time graph of an object is shown.

Use this graph to:

(a) find the final displacement of the object,

(b) draw the equivalent acceleration/time graph.

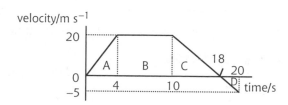

Answer to (a)

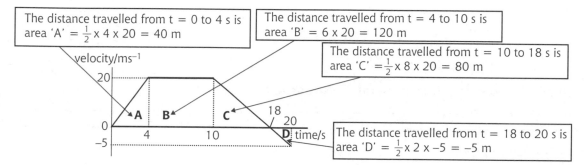

The distance travelled from t = 0 to 4 s is area 'A' = $\frac{1}{2}$ x 4 x 20 = 40 m

The distance travelled from t = 4 to 10 s is area 'B' = 6 x 20 = 120 m

The distance travelled from t = 10 to 18 s is area 'C' = $\frac{1}{2}$ x 8 x 20 = 80 m

The distance travelled from t = 18 to 20 s is area 'D' = $\frac{1}{2}$ x 2 x –5 = –5 m

The total displacement = 40 + 120 + 80 – 5 = 235 m

Answer to (b)

From t = 0 to 4 s, the acceleration = (v – u)/t = 20/4 = 5 m s^{-2}.

From t = 4 to 10 s, the acceleration = (v – u)/t = 0/6 = 0 m s^{-2}.

From t = 10 to 18 s, the acceleration = (v – u)/t = –20/8 = –2.5 m s^{-2}.

From t = 18 to 20 s, the acceleration = (v – u)/t = –5/2 = –2.5 m s^{-2}.

This gives the following graph.

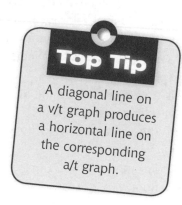

Top Tip

A diagonal line on a v/t graph produces a horizontal line on the corresponding a/t graph.

Quick Test 5

1. Which of these graphs show(s) constant velocity?

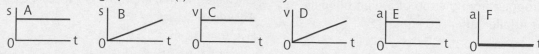

2. The velocity/time graph for an object is;

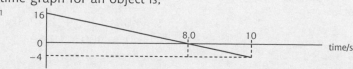

Calculate (a) the acceleration

(b) the displacement after 10 s.

The Equations of Motion

Equations, Symbols and Units

The equations of motion are relationships between displacement, velocity, acceleration and time.

The equations

$$s = \frac{1}{2}(u + v)t$$

$$s = ut + \frac{1}{2}at^2$$

$$v = u + at$$

$$v^2 = u^2 + 2as$$

The symbols and units

s = displacement (m)

u = initial velocity (m s^{-1})

v = final velocity (m s^{-1})

a = acceleration (m s^{-2})

t = time (s)

These equations can only be used when the **acceleration is constant** (or 'uniform').

Choosing appropriate equations

To choose the appropriate equation to be used in a question:
- make a list of the three quantities given in the question. (Some of these quantities may be 'hidden'. For example when an object is 'dropped' you need to list the initial velocity, u, as 0 m s^{-1} and the acceleration, a, as 9.8 m s^{-2}.);
- add to this list the quantity you have been asked to find.

This list of four quantities then indicates which one of the equations should be used.

Top Tip

If you choose the upward velocity to be positive, you must make the acceleration due to gravity negative.

Example 1
An car is travelling with an initial velocity of 3.0 m s^{-1}.
It now accelerates at 2.0 m s^{-2}.
Calculate the value of its velocity 8.0 s later.

Answer

List: $u = 3.0$ m s^{-1} $a = 2.0$ m s^{-2} $t = 8.0$ s $v = ?$

Equation: $v = u + at$

Substitution and solution:

$\Rightarrow v = 3.0 + (2.0 \times 8.0)$

$\Rightarrow v = 19$ m s^{-1}

Example 2
A brick is dropped from the top of a high building.
How far has it fallen after 3.0 s?

Answer

List: $u = 0$ m s^{-1} $a = 9.8$ m s^{-2} $t = 3.0$ s $s = ?$

Equation: $s = ut + \frac{1}{2}at^2$

Substitution and solution:

$\Rightarrow s = (0 \times 3) + (\frac{1}{2} \times 9.8 \times 3^2)$

$\Rightarrow s = 0 + \left(\frac{1}{2} \times 9.8 \times 9\right)$ m

$\Rightarrow s = 44.1$ m

Example 3
An object is fired vertically upwards with an initial speed of 30 m s^{-1}.
Calculate the maximum height it reaches.

Answer

List: $u = 30$ m s^{-1}; $v = 0$; $a = -9.8$ m s^{-2}; $s = ?$

Equation: $v^2 = u^2 + 2as$

Substitution and solution:

$\Rightarrow 0^2 = 30^2 + 2(-9.8)s$

$\Rightarrow 0 = 900 - 19.6s$

$\Rightarrow 19.6s = 900$

$\Rightarrow s = 45.92 = 46$ m

Example 4

A vehicle accelerates uniformly from a speed of 5.6 m s⁻¹ to a speed of 24.4 m s⁻¹ in a time of 8.5 s. How far does it travel in this time?

Answer

List: $u = 5.6$ m s⁻¹; $v = 24.4$ m s⁻¹; $t = 8.5$ s; $s = ?$

Equation: $s = \frac{1}{2}(u + v) t$

Substitution and solution:

$$\Rightarrow s = \frac{1}{2}(5.6 + 24.4) \times 8.5$$

$$\Rightarrow s = \frac{1}{2}(30) \times 8.5$$

$$\Rightarrow s = 127.5 \text{ m}$$

Deriving the equations of motion

1. $\boxed{s = \frac{1}{2}(u + v)t}$

In words; displacement = average velocity × time = $\dfrac{(\text{initial velocity} + \text{final velocity})}{2} \times \text{time}$

In symbols; $s = \dfrac{(u + v)}{2} \times t = \frac{1}{2}(u + v)\, t$

2. $\boxed{v = u + at}$ This equation may be derived by rearranging the formula used to define acceleration.

$a = (v - u)/t$ Multiplying both sides by time t gives $at = (v - u)$.

Changing left- and right-hand sides gives $(v - u) = at$.

Adding u to both sides gives $\underbrace{(v - u)}_{= 0} + u = at + u$ OR $v = u + at$.

3. $\boxed{s = ut + \frac{1}{2}at^2}$ An object has an initial velocity, u. It accelerates with an acceleration, a, for a time, t. Its velocity/time graph is:

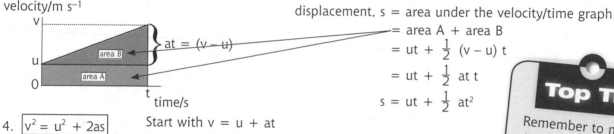

velocity/m s⁻¹

displacement, s = area under the velocity/time graph

= area A + area B

$= ut + \frac{1}{2}(v - u)\, t$

$= ut + \frac{1}{2}\, at \cdot t$

$s = ut + \frac{1}{2}\, at^2$

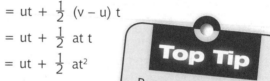

Top Tip

Remember to make a list of known values, plus the value to be found – then choose the appropriate equation.

4. $\boxed{v^2 = u^2 + 2as}$

Start with $v = u + at$

Square both sides $\Rightarrow$

$\Rightarrow v^2 = (u + at)(u + at)$

$\Rightarrow v^2 = u^2 + uat + uat + a^2t^2$

$\Rightarrow v^2 = u^2 + 2uat + a^2t^2$

$\Rightarrow v^2 = u^2 + 2a\underbrace{\left(ut + \frac{1}{2}\, at^2\right)}_{\text{equals displacement, s}}$

$\Rightarrow v^2 = u^2 + 2as$

Quick Test 6

1. A car starts from rest and accelerates at 2.0 m s⁻² for 10 seconds. Calculate the distance it travels in this time.

2. A girl pedals her bicycle so that it accelerates uniformly from rest to a speed of 10 m s⁻¹ over a distance of 50 m. Calculate her acceleration.

3. A boy accidentally drops his camera from the Eiffel Tower, which is 300 m high.
   (a) How long does it take for the camera to reach the ground?
   (b) How fast is the camera travelling just before it hits the ground?

4. A ball is thrown vertically into the air with a speed of 16 m s⁻¹. How long does it take to return to its launch point?

5. The brakes of a train are applied causing it to slow down from 28 m s⁻¹ to 6.0 m s⁻¹ in a time of 44 s. How far does it travel in this time?

Projectiles

A projectile is any object which has been thrown or fired into the air and so is then affected only by gravity and air resistance. Often the effects of air resistance can be ignored so that only the effect of gravity needs to be considered. Gravity causes a force (called 'weight') which acts vertically downwards. **Gravity cannot affect horizontal velocity** and so the horizontal velocity of a projectile remains constant (when air resistance is ignored). Weight causes a uniform, vertical acceleration (= 9.8 m s^{-2} on Earth). The vertical velocity of a projectile therefore changes by 9.8 m s^{-1} each second (slowing down by 9.8 m s^{-1} each second on its way up and speeding up by 9.8 m s^{-1} each second on its way down).

Weight

Time to maximum height

The time to reach maximum height is determined by the initial **vertical component** of velocity. Maximum height is reached by a projectile when its vertical component of velocity has decreased to zero. On Earth, the vertical component of velocity changes by 9.8 m s^{-1} each second giving;

$v = u + at$

$0 = u + (-9.8)\,t \quad \Rightarrow \quad t = u/9.8$ or $\boxed{\text{time to maximum height} = \dfrac{\text{initial vertical velocity}}{9.8}}$

Time of flight
Ignoring air resistance, a projectile takes the same time to fall back to the point of launch as it took to rise to its maximum height. Therefore

$\boxed{\text{total time of flight} = 2 \times \text{time to max height}}$

Range
The range of a projectile means the horizontal distance, R, travelled until it hits the ground again.
Ignoring the effects of air resistance, the horizontal speed does not change and so this distance is

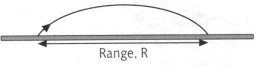

Range, R

$\boxed{\text{horizontal range} = \text{horizontal component of velocity} \times \text{total time of flight}}$

Example
A projectile is fired with a speed of 55 m s^{-1} at an angle of 24° to the horizontal. Find
(a) its initial horizontal speed
(b) its horizontal speed after 4.0 s
(c) its initial vertical speed
(d) its vertical speed after 2.0 s
(e) the time to reach maximum height
(f) its horizontal range.

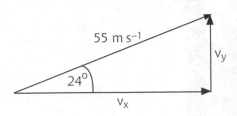

Top Tip

You must always resolve the initial velocity of any projectile into its horizontal and vertical components (see page 8).

Answer
(a) $v_x = v \cos \theta = 55 \cos 24° = 50.2$ m s^{-1}
(b) 50.2 m s^{-1} (because it stays constant)
(c) $v_y = v \sin \theta = 55 \sin 24° = 22.4$ m s^{-1}
(d) after 2.0 s, $v_y = 22.4 - (2 \times 9.8) = 2.8$ m s^{-1} upwards
(e) List: $u = 22.4$ m s^{-1}; $a = -9.8$ m s^{-2}; $v = 0$; $t = ?$
   Equation of motion: $v = u + at$
   Substitution: $\Rightarrow 0 = 22.4 + (-9.8)\,t$
   $\Rightarrow t = (0 - 22.4)/-9.8 = 22.4/9.8 = 2.29$ s
(f) total time of flight $= 2 \times$ time to max. height $= 2 \times 2.29 = 4.58$ s
   range $=$ horiz. speed $\times$ time of flight $= 50.2 \times 4.58 = 230$ m

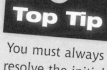

55 m s^{-1}

v_y

24°

v_x

Free fall and terminal velocity

Near the surface of the earth all objects fall with a downward acceleration of 9.8 m s⁻², when air resistance is ignored. In many situations, however, it is not realistic to ignore air resistance. One example is the free-fall parachutist. As the parachutist falls faster, the upward acting force of air resistance increases until, eventually, this force balances the weight of the parachutist. As a result the speed becomes constant at a value called the **terminal velocity**.

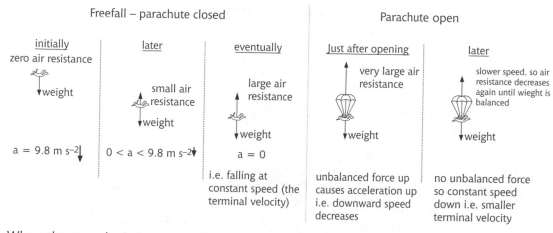

Freefall – parachute closed

initially	later	eventually
zero air resistance		
↓weight	small air resistance ↓weight	large air resistance ↓weight
$a = 9.8$ m s⁻²	$0 < a < 9.8$ m s⁻²	$a = 0$
		i.e. falling at constant speed (the terminal velocity)

Parachute open

Just after opening	later
very large air resistance	slower speed, so air resistance decreases again until wieght is balanced
↓weight	↓weight
unbalanced force up causes acceleration up i.e. downward speed decreases	no unbalanced force so constant speed down i.e. smaller terminal velocity

When the parachute is eventually opened, its large area produces an even greater air resistance and the resultant upward force causes deceleration until the forces are again balanced at a **lower** terminal velocity. The velocity/time graph for the fall is shown below.

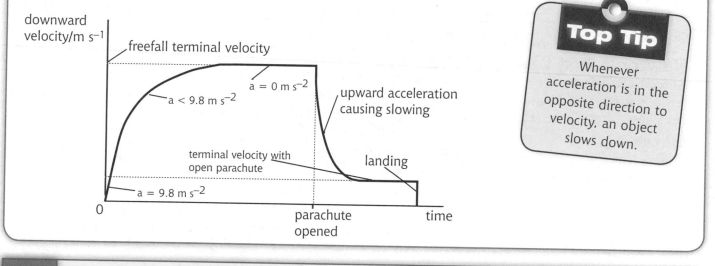

Top Tip

Whenever acceleration is in the opposite direction to velocity, an object slows down.

Quick Test 7

1. An object is fired into the air. The vertical component of its initial velocity is 36 m s⁻¹.
   Calculate: (a) the time taken for it to reach maximum height;
   (b) the maximum height reached by the object.

2. An object is fired from the ground at a speed of 49.0 m s⁻¹ at an angle of 30° to the horizontal.
   Calculate: (a) the horizontal component of its initial velocity;
   (b) the vertical component of its initial velocity;
   (c) the time taken to reach maximum height;
   (d) the maximum height reached by the object;
   (e) the horizontal distance travelled before it hits the ground again.

3. How would each of the answers to question 2 be affected when air resistance is not ignored?

Newton's Laws of Motion

Balanced and unbalanced forces

An object has **balanced forces** acting on it when they are **equal** and **opposite**.
Examples of objects with balanced forces.

(a)
8 N ← ☐ → 8 N

(b)
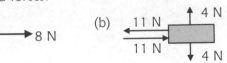

An object has an **unbalanced force** acting on it when the forces acting do **not** cancel out.
Examples of objects with unbalanced forces.

(a)
8 N ← ☐ → 7 N

Unbalanced force, F = 1 N acting to the left.

(b)

Unbalanced force, F = 2 N acting downwards.

The Three Laws of Motion

Sir Isaac Newton stated three laws about forces affecting the motion of objects.

Newton's First Law of Motion
Newton's **First Law** of Motion (what happens when **balanced** forces act on an object) is 'an object remains at rest or moving at constant speed in a straight line unless acted on by an unbalanced force'. In other words, an unbalanced force is needed to make an object speed up, slow down or change direction.

Example
A ship is travelling at a constant speed of 9.0 m s⁻¹ due east. Its engines produce a driving force of 65 000 N. What is the total resistive force acting on the ship?

Answer
Newton's First Law tells us that constant speed means that balanced forces act on the ship. The resistive force must therefore be equal in size to the driving force (= 65 000 N) and must act in the opposite direction. Total resistive force = 65 000 N acting due west. (Note that the value of the constant speed is an irrelevant piece of information).

Newton's Second Law of Motion
Newton's **Second Law** of Motion (what happens when **unbalanced** forces act on an object) is 'the acceleration of an object is directly proportional to the unbalanced force and inversely proportional to its mass.' In other words:
  • doubling the value of the unbalanced force doubles the acceleration, and
  • doubling the mass of the object halves the acceleration.
We usually use this relationship in the form of the formula to the right. It is essential to understand that 'F' is not just 'force' – it is the unbalanced force acting on the object.

Top Tip

'F' is the **unbalanced** force.

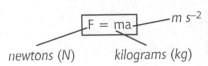

$$F = ma$$

newtons (N)   kilograms (kg)   m s⁻²

Example

A mass of 6.0 kg has a force of 7.0 N acting on it due west and also a force of 4.0 N acting due east. What is the resulting acceleration of the object?

Answer

The unbalanced force acting on the object is 3.0 N due west. So, acceleration, a = F/m = 3.0/6.0 = 0.5 m s^{-2} due west. (The direction of acceleration is always the same as the direction of the unbalanced force, 'F')

> Newton's Second Law leads to the definition of the newton: 'one newton is the size of the unbalanced force which causes a mass of one kilogram to accelerate at one metre per second squared.'

Newton's Third Law of Motion

Newton's **Third Law** (forces always occur in equal and opposite pairs) is 'when object A exerts a force on object B, then object B exerts an **equal** and **opposite** force on object A.'

Examples

(a) Kicking a football

    Force A = force forwards by foot on football

    Force B = force backwards by football on foot

(b) Car accident

    Force A = force by car on pedestrian (hence need for hospital treatment).

    Force B = force by pedestrian on car (hence need for garage treatment).

(c) Horse pulling a cart

    Force A = force forward by horse on cart

    Force B = force backwards by cart on horse

(d) Rocket

    Force A = force backwards by engines on exhaust gases

    Force B = force forward by exhaust gases on engine

A ← [ ] → B

(e) Weight

    Force A = force of attraction of earth on object

    Force B = force of attraction of object on earth

weight
reaction

Top Tip

These equal and opposite forces do not cancel out because they act on different objects.

Quick Test 8

1. What extra force(s) are required to ensure that this object's velocity is constant?
2. The frictional forces acting on a car are greater in magnitude than the driving force of the engine. Describe the motion of the car.
3. A van of mass 2,400 kg is on a 12° slope.
   The driving force from the engine is 6500 N.
   The total frictional forces are 3400 N.
   Calculate:
   (a) the component of the van's weight parallel to the slope;
   (b) the unbalanced force acting on the van;
   (c) the acceleration of the van.
4. A book of mass 0.75 kg is at rest on a table.
   What is the size of the reaction force the table exerts on the book?

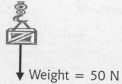

Weight = 50 N

m g sin ϑ

Driving force = 6500 N

Friction = 3400 N

12°

Work, Energy and Power

Work, E_w

Whenever a force, F, causes an object to move through a distance, d, **work** is done. This means that there has been a **transfer of energy**. For example, when a gardener pushes a wheelbarrow, some energy is transferred from the person's body and becomes kinetic energy in the wheelbarrow. The formula used to calculate the quantity of energy transferred is shown below.

work done $=$ force $\times$ distance   or, in symbols $E_w = F\,d$

joules   *newtons*   *metres*   *J*   *N*   *m*

The units of work are newton metres (N m) or joules (J). Work is a **scalar** quantity.

Example
A force of 260 N is used to push a box of mass 15 kg for a distance of 3.7 m along a horizontal surface. Calculate the work done on the box.

Answer
$E_w = F\,d = 260 \times 3.7 = 962$ J. (The value of the mass of the box is irrelevant.)
If the horizontal surface is frictionless, all of this energy is transferred into the kinetic energy of the box.
If there are frictional forces acting (much more likely!), most of the energy becomes heat energy.

Kinetic energy, E_k

This is the energy an object has when it is **moving**. The quantity of kinetic energy an object has depends on the **mass** of the object and how fast it is moving (i.e. the object's **velocity**).

The formula for calculating kinetic energy is

kinetic energy $= \frac{1}{2} \times$ mass $\times$ velocity^2 or, in symbols $E_k = \frac{1}{2}\,m\,v^2$

joules   *kilograms*   *metres per second*   *J*   *kg*   *m s^{-1}*

The units of kinetic energy are joules (J). E_k is a **scalar** quantity.

Top Tip

You must remember to square the value of the velocity.

Example
A racing car of mass 600 kg is travelling at 80 m s^{-1}. Calculate its kinetic energy.

Answer
$E_k = \frac{1}{2}\,m\,v^2 = \frac{1}{2} \times 600 \times 80^2 = 1.92 \times 10^6$ J.

Gravitational potential energy, E_p

This is the energy an object has because it is **at a height**. The quantity of potential energy an object has depends on the **mass** of the object, the gravitational field strength, **g**, and its **vertical height**. The formula used for calculating potential energy is shown below.

potential energy $=$ mass $\times$ g $\times$ height   or, in symbols $E_p = m\,g\,h$

joules   *kilograms*   *N kg^{-1}*   *metres*   *J*   *kg*   *N kg^{-1}*   *m*

The units of potential energy are joules (J). E_p is a **scalar** quantity.

Example

An athlete of mass 75 kg is at a height of 2.2 m. Calculate his potential energy.

Answer

$E_p = m\,g\,h = 75 \times 9.8 \times 2.2 = 1617$ J $= 1600$ J. (The final answer has been rounded to two significant figures to match the given data.)

Potential energy to kinetic energy (and vice versa)

When an object moves closer to the centre of the Earth, it loses potential energy (even when it does not move vertically – for example, as it slides down a slope). As a result, there is a gain in kinetic energy. In the absence of frictional forces, the **gain** in kinetic energy is equal to the **loss** of potential energy,

i.e. $\boxed{\dfrac{1}{2}\,m\,v^2 = m\,g\,h}$

This relationship can be rearranged to give $\boxed{v = \sqrt{2\,g\,h}}$.

This formula can be used (in the absence of frictional forces) to find either:

- the final velocity, v, of an object dropped through a vertical height, h; or
- to find the height, h, reached by an object launched with an initial vertical velocity, v.

It is important to note that this relationship ($v = \sqrt{2\,g\,h}$) is **not** listed in the Physics data booklet.

Power, P

Power is the **rate** of doing work (i.e. the work done per second). A high-powered machine is able to do a lot of work in a short time. For example, a more powerful car has a greater acceleration than a lower-powered car. (It is also able to reach a higher top speed and pull a heavier load.)

The formula used to calculate power is

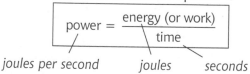

$$\text{power} = \frac{\text{energy (or work)}}{\text{time}}$$

or, in symbols

$$P = \frac{E}{t} \begin{array}{l} J \\ s \end{array}$$

joules per second   *joules*   *seconds* $J\,s^{-1}$

Top Tip

Using J s⁻¹ for the units of power can help you remember the meaning of 'power'.

The units of power are joules per second ($J\,s^{-1}$) or watts (W).   Power is a **scalar** quantity.

Example

A lift can carry a load of twelve passengers of average mass 85 kg through a height of 20 m in 15 s. Calculate the minimum power output of the lift's motor.

Answer

Total mass lifted = $12 \times 85 = 1020$ kg.

Work done = potential energy gained = $m\,g\,h = 1020 \times 9.8 \times 20 = 199\,920$ J

$P = E/t = 199\,920/15 = 13328 = 13$ kJ s^{-1} or 13 kW.

Quick Test 9

1. For each of the quantities work, energy and power, give the units and state whether the quantity is a scalar or a vector.
2. An object is moved through a distance of 3.5 m by applying a force of 96 N. Calculate the work done.
3. A coach of mass 12 500 kg is travelling at 32 m s⁻¹ on a motorway. Calculate its kinetic energy.
4. A car of mass 1 400 kg is travelling at 24 m s⁻¹. The driver now applies the brakes and the car's speed reduces to 8.0 m s⁻¹. Calculate the loss in the car's kinetic energy.
5. An object falls to the ground from a height of 15 m. Calculate its speed just before it lands.
6. A device can do 8.4 kJ of useful work in 2 minutes. Calculate its output power in watts.

Collisions and Momentum

Momentum

There are many collisions occurring all the time in the world around us – for example vehicles on roads, snooker balls against each other, atoms and molecules. Physics can analyse and predict the outcome of collisions using the quantity called **momentum**.

| momentum = mass × velocity | or, in symbols | $p = m\,v$ |

           kilograms   *metres per second*

The units of momentum are therefore equal to the units of mass multiplied by the units of velocity, i.e. **kilogram metres per second** (kg m s^{-1}). The reason why momentum is useful in analysing collisions is because experiments show that the total momentum of a system remains constant before and after a collision. This is known as the law of **conservation of momentum** and is usually written as, 'total momentum before a collision = total momentum after a collision, in the absence of external forces.' Momentum is a **vector** quantity – if an object moving to the right is said to have a positive momentum, then an object moving to the left has a negative momentum.

Example 1
A 3.0 kg object is travelling at 5.0 m s^{-1}. It collides with a stationary mass of 2.0 kg. They stick together. What is their common velocity after the collision?

Answer

Before   3.0 kg → 5.0 m s^{-1} 2.0 kg 0 m s^{-1}

After   5.0 kg   v = ? m s^{-1}

          Total momentum before = total momentum after

          $m_1u_1 + m_2u_2 = mv$

$\Rightarrow$ $(3.0 \times 5.0) + (2.0 \times 0) = 5.0\,v$

$\Rightarrow$ $15 = 5\,v$

$\Rightarrow$ $v = 3.0$ m s^{-1}

i.e. they carry on with a speed of 3.0 m s^{-1} in the same direction.
Note that the changes in momentum for each object are equal in size (in this case, 6.0 kg m s^{-1}) and opposite in direction – this is true in all collisions.

Example 2
A 4.0 kg mass travelling to the right at 4.0 m s^{-1} collides with a 5.0 kg mass which is travelling to the left at 2.0 m s^{-1}. After the collision the 4.0 kg mass bounces backwards at 1.0 m s^{-1}. What does the 5.0 kg mass do?

Answer

Before   4.0 kg → 4.0 m s^{-1} 2.0 m s^{-1} ← 5.0 kg

After   1.0 m s^{-1} ← 4.0 kg   5.0 kg v = ? m s^{-1}

Total momentum before = total momentum after

$$m_1u_1 + m_2u_2 = m_1v_1 + m_2v_2$$
$$\Rightarrow \quad (4.0 \times 4.0) + (5.0 \times -2.0) = (4.0 \times -1.0) + 5v$$
$$\Rightarrow \quad 16 - 10 = -4 + 5v$$
$$\Rightarrow \quad 5v = 10$$
$$\Rightarrow \quad v = 2.0 \text{ m s}^{-1}$$

i.e. the 5.0 kg object moves at a speed of 2.0 m s^{-1} to the right.

Top Tip

The negative signs are essential because velocity is a vector quantity.

Types of collisions

In all collisions, the total momentum remains constant.
In all collisions, the total energy (i.e. all forms of energy added together) remains constant.

For Higher Physics you need to be able to identify whether a collision is **elastic** or **inelastic**. The type of collision is determined by what happens to the total kinetic energy, E_k. In an **elastic** collision the total momentum remains constant and the total kinetic energy, E_k, also remains constant. In an **inelastic** collision the total momentum remains constant but the total kinetic energy, E_k, does not remain constant.

Example 3

A mass of 3.0 kg is travelling at 4.0 m s^{-1} to the right. It collides with a mass of 2.0 kg which is travelling at 2.0 m s^{-1} in the same direction. After the collision, the 3.0 kg object is moving at 3.0 m s^{-1} to the right and the 2.0 kg object is moving at 3.5 m s^{-1} to the right. What type of collision is this?

Top Tip

Deciding the type of collision is **not** about whether the objects bounce apart.

Answer

Before   3.0 kg ⟶ 4.0 m s^{-1} 2.0 kg ⟶ 2.0 m s^{-1}

After   3.0 kg ⟶ 3.0 m s^{-1} 2.0 kg ⟶ 3.5 m s^{-1}

Total E_k before $= \frac{1}{2} m_1 u_1^2 + \frac{1}{2} m_2 u_2^2 = \frac{1}{2}\, 3.0 \times 4.0^2 + \frac{1}{2}\, 2.0 \times 2.0^2 = 24 + 4 = 28$ J

Total E_k after $= \frac{1}{2} m_1 v_1^2 + \frac{1}{2} m_2 v_2^2 = \frac{1}{2}\, 3.0 \times 3.0^2 + \frac{1}{2}\, 2.0 \times 3.5^2 = 13.5 + 12.25 = 25.75$ J

The total kinetic energy has decreased (from 28 J to 26 J).
The collision is therefore not elastic – it is inelastic.

Quick Test 10

1. Write out the formula for calculating momentum
   (a) in words, and
   (b) as symbols.
2. What are the units for momentum?
3. Is momentum a scalar quantity or a vector quantity?
4. A 6.0 kg object is travelling at 2.0 m s^{-1}. It collides with a stationary mass of 2.0 kg. They stick together. What is their common velocity after the collision?
5. A 6.0 kg mass and an 8.0 kg mass are held together stationary, with a squashed spring between them. When released, the 8.0 kg travels to the right at 4.0 m s^{-1}.
   (a) What does the 6.0 kg do?
   (b) Is this interaction elastic?

Before   6.0 kg ▬ 8.0 kg

spring

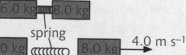

After   v = ?m s^{-1} 6.0 kg ⟲⟲⟲ 8.0 kg ⟶ 4.0 m s^{-1}

Impulse

Force on an object

When a force acts on an object, the effect that this force has on the object depends on two factors:
(a) the size of the force, and
(b) how long a time the force acts for. These factors are combined in the quantity known as **impulse**.

| impulse = force × time |

| impulse = F t |

newtons   seconds

The units of impulse are newton seconds (N s).   Impulse is a **vector** quantity.

Example
A golf club exerts a force of 4000 N for a time of 3.0 ms on a golf ball.
What is the impulse on the golf ball?
Answer
Impulse = F × t
$$= 4000 \times 3 \times 10^{-3}$$
$$= 12 \text{ N s}$$

Top Tip

The units of impulse are **not** newtons per second.

The connection between impulse and momentum

Starting from Newton's Second Law,
$$F = m\,a$$
$$\Rightarrow \quad F = m(v - u)/t$$
Multiply both sides by time, t
$$\Rightarrow \quad F t = m(v - u)$$
$$\Rightarrow \quad \boxed{F t = mv - mu}$$

| Impulse = change in momentum |

impulse   final momentum   initial momentum

The units of impulse and momentum must therefore be equivalent, i.e.
newton-seconds = kilogram metres per second (i.e. 1 N s = 1 kg m s^{-1}).
When two objects collide,
change in momentum of A = − change in momentum of B.
Dividing both sides of this relationship by the time of contact, t
$$\Rightarrow \quad (\text{change in momentum of A})/t = -(\text{change in momentum of B})/t$$
$$\Rightarrow \quad \text{force on object A} = -\text{force on object B (i.e. this is Newton's Third Law – see page 19)}.$$

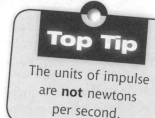

Impulse and force/time graphs

| Impulse is equal to the area under a force/time graph |

Example
An object of mass 3.0 kg is initially moving at 0.5 m s^{-1}.
It is then acted on by a force as shown in the following graph.

Force/N
80

0   40   140   time/ms

Calculate the final speed of the object.

Answer

Impulse = area under the force/time graph

$$= \frac{1}{2} \times 140 \times 10^{-3} \times 80$$

$$= 5.6 \text{ N s}$$

Change in momentum = impulse = 5.6

$\Rightarrow \quad mv - mu = 5.6$

$\Rightarrow \quad (3 \times v) - (3 \times 0.5) = 5.6$

$\Rightarrow \quad 3v = 5.6 + 1.5 = 7.1$

$\Rightarrow \quad$ final velocity, $v = 7.1/3 = 2.37 \text{ m s}^{-1}$

Experiment

Experiment to find the average force exerted on a football.

Apparatus

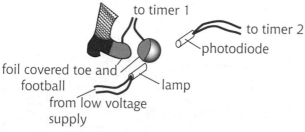

Top Tip

The principles used in this experiment can be applied to find the average force on any object during any collision.

Procedure

Timer 1 records the time of contact between the foot and the ball thanks to an electrical circuit being completed when the foot and ball are touching each other.

Timer 2 records the length of time the diameter of the ball takes to pass through the light gate – from this, the speed of the ball after the kick, v, can be calculated.

Impulse = Ft = mv – mu

$\Rightarrow \quad F = (mv - mu)/t$

Now u = 0 because the ball was stationary before the kick,

$\Rightarrow \quad F = mv/t$

i.e. average force on ball $= \dfrac{\text{(mass of ball} \times \text{final velocity)}}{\text{time of contact}}$

Note that the actual force exerted by the foot on the ball will increase from zero, go through a maximum and then decrease to zero, i.e.

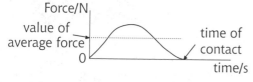

The area under this force/time graph will still be the impulse on the ball (and equal to its change in momentum).

Quick Test 11

1. (a) Write out the formula which gives the meaning of 'impulse'.
   (b) What are the units for impulse?
   (c) Is impulse a scalar quantity or a vector quantity?
2. A force of 16 N acts on an object of mass 1.7 kg for a time of 52 s. Calculate the impulse on the object.
3. A force of 75 N acts on an object of mass 6.9 kg for a time of 18 s. Calculate the object's change in momentum.
4. An object of mass 4.5 kg is travelling with an initial velocity of 1.3 m s^{-1}. It is now acted on by a force and receives an impulse of 36 N s. Calculate its final velocity.
5. An object is acted on by a force as shown by the graph. Calculate the object's change in momentum.

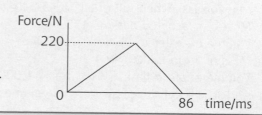

Density

You are likely to have heard the trick question, 'Which is heavier, a tonne of coal or a tonne of feathers?' The answer is 'neither' because they are the same mass and so have the **same weight** as each other. The difference between them is that the feathers take up a lot more space – the volume of the feathers is much greater. The volume of the coal is less because its atoms are more closely packed together. We say that the coal has a greater **density**. When a substance has a large mass concentrated into a small volume, we say it has a high density.

Formula defining density

$$\text{density} = \frac{\text{mass} \quad kg}{\text{volume}}$$
$$m^3$$

$$\rho = \frac{m}{v}$$

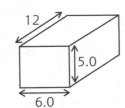

Density means 'mass per unit volume'.
The units of density are kilograms per cubic metre (kg m⁻³).   Density is a scalar quantity.
Note that there are 1,000 grams in 1.0 kg and 1 000 000 cm³ in 1.0 m³.

Example
An object measures 5.0 cm × 6.0 cm × 12 cm. It has a mass of 500 g. What is its density?

Answer
Volume = $5 \times 6 \times 12$ = 360 cm³ = 0.00036 m³.
Density = mass/volume = 0.5/0.00036 = 1389 kg m⁻³.

To find the density of a substance or object

- Measure the mass of the object (e.g. using an electronic balance).
- Measure the volume of the object (e.g. using a measuring cylinder for a liquid).
- Calculate the density using the formula.

Example
The density of air – one possible experimental arrangement.

Apparatus: strong glass flask
measuring cylinder
vacuum pump
sensitive electronic balance

Safety Tip

The use of a safety screen and safety glasses is recommended. Even a strong glass flask could implode and spray out dangerous glass fragments.

Procedure
(i)   The mass of the flask is measured both before and after being evacuated – the difference in these masses is the mass of the air in the flask.
(ii) The flask is filled with water which is then poured into the measuring cylinder – the volume of this water is equal to the volume of the air which had been in the flask.
(iii) These values of mass and volume are substituted into the density formula to give the density of air.

Typical results and calculations:
Mass of flask when full of air = 454.4 g
Mass of evacuated flask = 453.1 g
Volume of water to fill flask = 1.0 litre
Mass of air in flask = 454.4 − 453.1 = 1.3 g = 0.0013 kg
Volume of air in flask = 1.0 litre = 1,000 cm³ = 0.001 m³
And so the density of air = mass/volume = 0.0013/0.001 = 1.3 kg m⁻³ (the accepted value is 1.29 kg m⁻³)

Densities and particle arrangements

	Substance	Density (kg m⁻³)	Particle picture

Solids

	Substance	Density (kg m^{-3})
	aluminium	2700
	iron	7900
	gold	19 300
	carbon (graphite)	2300
	glass (different types)	2400 to 4800

Liquids

	water	1000
	glycerine	1300
	methylated spirit	800

Gases

	hydrogen	0.09
	helium	0.18
	oxygen	1.43
	nitrogen	1.25
	carbon dioxide	1.98

Comparisons between solids, liquid and gases

Generally, solids and liquids have high values of density and gases have low values. The density of a substance in its solid form is approximately the **same** as its density in its liquid form. This is because the **distance between the molecules** is about the same in the solid state as in the liquid state (and is approximately equal to one molecular diameter). When a substance changes to its gaseous state, its volume **increases** by approximately 1,000 times (and so its density **decreases** by about 1,000 times). This means that in the gaseous state the molecules are, on average, about 10 diameters apart.

Top Tip

Remember that there is little difference between the values when a solid changes to a liquid – the big changes occur when it becomes a gas.

Summary

	Solid	Liquid	Gas
Relative Volume	1	1	1,000
Relative Density	1,000	1,000	1
Molecular Distance	1	1	10

1. (a) Write out the formula which gives the meaning of 'density'.
   (b) What are the units for density?
   (c) Is density a scalar quantity or a vector quantity?
2. A substance has a mass of 5 600 g and a volume of 0.0020 m³. Calculate the density of the substance.
3. The volume of a sample of gold is 1.3 × 10⁻⁶ m³. Calculate the mass of the sample. (The density of gold can be found near the top of this page.)
4. An ice cube melts to become a small puddle of water. Explain what happens to the density.
5. A liquid evaporates and becomes a gas. Explain what happens to the density.

Pressure

High heels and elephants

High heels

A lady wearing high heels is more likely to cause damage to a tiled floor than an elephant. This is not because she is heavier than the elephant (!) but because her weight is concentrated on to the very small area of the heel. The elephant's weight is spread out over a much larger area. This effect of a force on a surface is embodied in the quantity 'pressure', which means the **force per unit area**.

Elephant's foot

Formula

$$\text{pressure} = \frac{\text{force} \quad \text{— } newtons}{\text{area} \quad \text{— } square\ metres}$$

The units of pressure are newtons per square metre (N m^{-2}) OR pascals (Pa).
1 Pa = 1 N m^{-2} (i.e. one pascal is the same as one newton per square metre).

Top Tip

There are 10 000 square centimetres in each square metre:
1 m^2 = 10 000 cm^2

Example

A box of mass 6.5 kg rests on the surface of a table.

The dimensions of the box are 0.50 m by 0.60 m by 1.2 m.

(a) What pressure does the box exert on the table when it rests on its largest area?

(b) What pressure does the box exert on the table when it rests on its smallest area?

Answer

Force on table   = weight of box = m g = 6.5 × 9.8 = 63.7 N

(a) Largest area   = 0.60 × 1.2 = 0.72 m^2

⇒ pressure = F/A = 63.7/0.72 = 88.5 N m^{-2}

(b) Smallest area = 0.50 × 0.60 = 0.30 m^2

⇒ pressure = F/A = 63.7/0.30 = 212.3 N m^{-2}

Pressure in gases

The molecules of a gas are in constant, random motion. This is called the **kinetic theory**.

The molecules are constantly colliding with the walls of their container. It is these collisions (between the molecules and the walls) which cause a force (and so a pressure) to be exerted by any gas on the walls of its container.

The pressure of a gas can be measured by an instrument called a Bourdon gauge.

The average pressure exerted by the air around us at sea level is 101,000 Pa (or 1.01 × 10^5 Pa) i.e. normal atmospheric pressure is 1.01 × 10^5 Pa. This pressure decreases as altitude increases. Weather forecasters use readings of atmospheric pressure as part of the information they need in order to predict the weather.

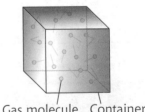

Gas molecule   Container

Bourdon gauge

Pressure in liquids

Pressure increases with increasing depth. This is why submarines need to have very strong hulls – the surrounding water causes a large crushing force.

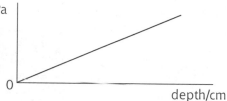

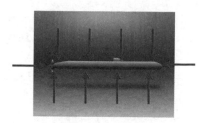

The graph has a straight line through the origin, proving that the extra pressure is **directly proportional** to **depth**, i.e. doubling the depth doubles the extra pressure.

The pressure under the surface of a liquid is also affected by the density of the liquid.

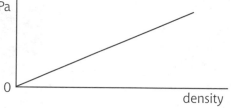

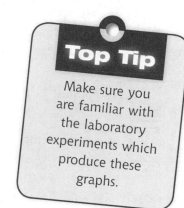

Top Tip

Make sure you are familiar with the laboratory experiments which produce these graphs.

The graph showing a straight line through the origin proves that the extra pressure is **directly proportional** to the **density** of the liquid.

The formula for the extra pressure under the surface of a liquid is:

Extra pressure = density of fluid × gravitational field strength × depth   or, in symbols $P = \rho\,g\,h$

$kg\,m^{-3}$ $N\,kg^{-1}$ m

Example

At a depth of 15.0 m in sea water the extra pressure is 1.50×10^5 Pa. Calculate

(a) the density of sea water; and

(b) the extra pressure at a depth of 24.0 m in sea water.

Answer

(a) $\rho = P/g\,h = 150\ 000/(9.8 \times 15.0) = 1.02 \times 10^3\ kg\,m^{-3}$

(b) $P = \rho\,g\,h = 1.02 \times 10^3 \times 9.8 \times 24.0 = 239\ 904 = 2.40 \times 10^5$ Pa

Quick Test 13

1. A surface of area 0.30 m² experiences a force of 690 N acting on it. Calculate the pressure on the surface.
2. A lady of weight 750 N is wearing high heels. At one moment all her weight is concentrated onto one of these heels which has an area of 1.5 cm². Calculate the pressure on the surface at that moment.
3. Explain why a gas exerts a pressure on the walls of its container.
4. List the factors which affect the extra pressure experienced under the surface of a liquid.
5. A diver is working at a depth of 56 m in sea water. Calculate the extra pressure due to the water (density of sea water = 1.02×10^3 kg m⁻³).

Upthrust and Flotation

When any object is immersed in a liquid, it experiences an upward force called 'upthrust'. Being a force, upthrust is measured in **newtons** and is a **vector** quantity.

For example, when a 500 g mass is suspended from a spring balance and then lowered into a beaker of water the reading on the balance **decreases**.

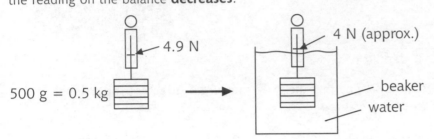

500 g = 0.5 kg

← 4.9 N

4 N (approx.)

beaker
water

This lower reading is not because the weight of the 500 g mass has decreased: it is because the water is exerting an **upward force** (of approximately 1 N) on the mass.

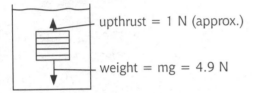

upthrust = 1 N (approx.)

weight = mg = 4.9 N

Top Tip

Be prepared to explain, in this much detail, how upthrust is produced.

The reason why a fluid (i.e. any liquid or gas) exerts an upthrust on an object immersed in it can be explained as follows:

- Pressure increases with depth, h, according to the relationship $P = \rho g h$.
- The pressure on the bottom surface, P_b, is therefore greater than the pressure on the top surface, P_t.
- Force, F, is directly proportional to the pressure, P, according to the relationship $F = P A$.
- The upward force on the bottom surface, F_b, is greater than the downward force on the top surface, F_t, because $P_b > P_t$.
- There is therefore an unbalanced force, F_U, acting upwards on the object – this unbalanced force is called upthrust.

Here is the explanation in pictures

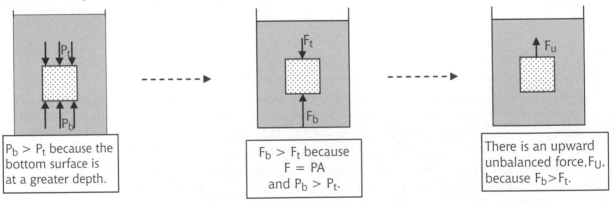

$P_b > P_t$ because the bottom surface is at a greater depth.

$F_b > F_t$ because
$F = PA$
and $P_b > P_t$.

There is an upward unbalanced force, F_U, because $F_b > F_t$.

You should note that because the upthrust force is due to the pressure **difference** between the top and bottom surfaces of the object, the upthrust is the **same at all depths** (assuming that the density of the liquid and the shape of the object do not change).

The upthrust force creates a buoyancy effect on any object in any fluid.

A ship floats when the water provides a large enough value of upthrust to balance the ship's weight. i.e.

Upthrust

Weight

Modern cruise ships can have masses as large as 50 000 000 kg. The weight of such a ship is 490 00 000 N. The upthrust required to keep it afloat is therefore also equal to 490 000 000 N!

However, an object sinks when its weight is greater than the upthrust.

This may be done intentionally – for example in a submarine. The ballast tanks of the submarine are allowed to fill with water. This causes the weight to increase and the submarine sinks. When compressed air is used to force water out of these ballast tanks, the weight of the submarine decreases. Once its weight becomes less than the upthrust, the submarine rises again.

Top Tip

Be prepared to explain floating and sinking in terms of the forces acting.

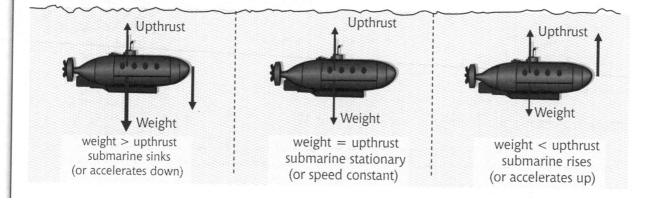

Upthrust

Weight

weight > upthrust
submarine sinks
(or accelerates down)

Upthrust

Weight

weight = upthrust
submarine stationary
(or speed constant)

Upthrust

Weight

weight < upthrust
submarine rises
(or accelerates up)

1. Give the units of upthrust and state whether it is a scalar or vector quantity.
2. A hot air balloon is rising at a constant speed.
   (a) Sketch a diagram showing all the forces acting on the balloon.
   (b) Write a relationship between these forces.
3. A ship of mass 3 600 000 kg is floating in a harbour.
   Calculate the upthrust exerted on the ship by the water.
4. A submarine of mass 2 400 000 kg experiences an upthrust of 25 000 000 N.
   Describe the vertical motion of the submarine.

The Gas Laws 1

The large scale, physical quantities which describe any sample of a gas are:
pressure, P   volume, V   temperature, T   mass, m
The gas laws are a series of relationships between these quantities.

Boyle's Law

Boyle's Law is the relationship between the pressure, P, and the volume, V, of a gas (the temperature and mass of the gas being kept constant).

Apparatus

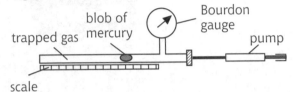

blob of mercury
Bourdon gauge
trapped gas
pump
scale

The mass is constant because the gas is trapped by the mercury – no gas molecules can enter or escape.

The temperature is constant because the apparatus is left for a short time before each reading is taken, to ensure that the gas settles to room temperature again.

The volume decreases as the pressure increases, producing the following graphs:

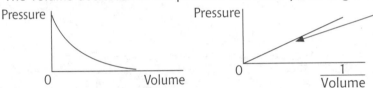

This straight line graph through the origin proves that the pressure is **inversely proportional** to volume (for T constant).

Top Tip

The data booklet does not list this formula. However, you may learn it and use it in the exam.

This relationship is often called Boyle's Law.

It can also be expressed as $\boxed{\text{pressure} \times \text{volume} = \text{constant}}$

This can be written as the following formula: $\boxed{P_1 V_1 = P_2 V_2}$

The law of pressures

The law of pressures is the relationship between the pressure, P, and temperature, T, of a gas (the volume and mass being kept constant).

Apparatus

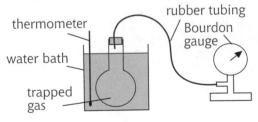

thermometer
rubber tubing
Bourdon gauge
water bath
trapped gas

The mass is constant because the system is sealed – no gas molecules can enter or escape from the flask.
The volume is constant because the rigid flask does not expand or contract significantly.

Pressure/Pa

0   100   Temperature/°C

The pressure increases as the temperature increases, as shown in the graph of the results.

Although this is a straight line, it does not pass through the origin. However, if this graph is extrapolated back to zero pressure, the **absolute zero** of temperature is found.

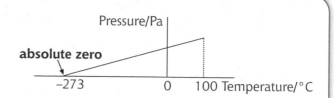

The Kelvin scale of temperature

When temperature is measured from absolute zero it is called the **Kelvin scale** of temperature. The size of the Kelvin degree is the same as the size of the Celsius degree; the two scales just start from different points, i.e.

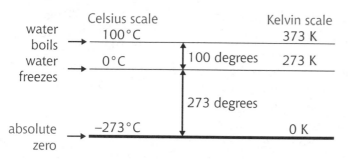

To change from °C to K, add 273.
To change from K to °C, subtract 273.
For example, normal human body temperature

$$= 37°C$$
$$= 37 + 273$$
$$= 310 K$$

The pressure/temperature graph can now be redrawn using temperatures in Kelvin.

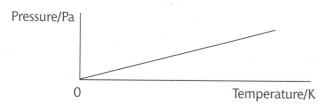

This straight line graph through the origin proves that pressure is **directly proportional** to temperature (but only when the temperature is in Kelvins, the mass is fixed and the volume is constant).

This relationship is known as the **law of pressures**.

This can also be expressed as:

| pressure/temperature = constant | (for temperature in Kelvins)

or written as the formula:

$$\frac{P_1}{T_1} = \frac{P_2}{T_2}$$ **so long as T is in Kelvins**

Top Tip

This formula is also worth learning even though it is not listed in the data booklet.

Quick Test 15

1. Sketch the graph which proves that the pressure of a gas is inversely proportional to its volume.
2. A gas occupies a volume of 36 cm³ when at a pressure of 1.0×10^5 Pa.
   What volume does it occupy when the pressure is changed to 2.5×10^5 Pa?
3. Change each of the following temperatures into their values on the Kelvin scale.
   (a) 0°C
   (b) 27°C
   (c) −23°C
   (d) 273°C
4. Sketch the graph which shows the relationship between the pressure of a gas and its temperature.
5. A sample of a gas is at a pressure of 1.01×10^5 Pa when it is in a room at a temperature of 22°C.
   What is the new pressure when the temperature is raised to 100°C?

The Gas Laws 2

Charles' Law

Charles' Law is the relationship between the volume, V, and the temperature, T, of a gas (the pressure and mass being kept constant).

Apparatus

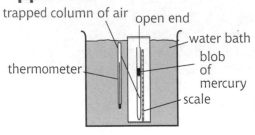

The mass is constant because the gas is trapped by the mercury – none can enter or escape.

The pressure is constant as the blob of mercury will move up or down until the pressure on its bottom surface balances the atmospheric pressure on its top surface.

Top Tip

You must change temperatures into their Kelvin values before substituting into any gas law.

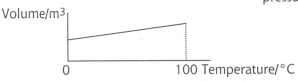

The volume increases as the temperature increases, as shown in the graph of the results.

As in the law of pressures, this graph can be extrapolated back to zero pressure. Again, this finds the absolute zero of temperature to be –273°C, as shown on the right.

Redrawing the above graph using temperatures in Kelvins gives this graph.

This graph of a straight line through the origin proves that the volume is directly proportional to temperature (but only when the temperature is in Kelvins, the mass is fixed and the volume is constant).

This relationship is known as Charles' Law.

It can also be expressed as

| volume/temperature = constant |

(for temperature in Kelvins)

or written as the formula

$$\frac{V_1}{T_1} = \frac{V_2}{T_2}$$ as long as T is in Kelvins

Top Tip

Another formula worth memorising!!

The General Gas Law

The three gas laws can be combined into one formula called the General Gas Law. This law allows calculations to be done when all three variables (P, V and T) change at the same time.

The formula is

$$\frac{P_1 V_1}{T_1} = \frac{P_2 V_2}{T_2}$$ as long as T is in Kelvins

Top Tip

This formula is listed in the data booklet. Make sure you practise using it.

Example

One litre of a gas is at a pressure of 1.5×10^5 Pa and a temperature of 25°C. The gas is now heated to a temperature of 120°C and the pressure is reduced to 0.8×10^5 Pa. Calculate the new volume.

Answer

$T_1 = 25°C = 25 + 273 = 298$ K; $T_2 = 120°C = 120 + 273 = 393$ K

$$\frac{P_1V_1}{T_1} = \frac{P_2V_2}{T_2} \Rightarrow \frac{1.5 \times 10^5 \times 1}{298} = \frac{0.8 \times 10^5 \times V_2}{393}$$

$$\Rightarrow V_2 = \frac{393 \times 1.5 \times 10^5 \times 1}{298 \times 0.8 \times 10^5}$$

$$\Rightarrow V_2 = 2.47 \text{ litres}$$

Note that the units of the answer are the same as the original units.

The kinetic model and the gas laws

The **kinetic theory** of matter describes the **particles** of all substances being in **constant**, **random** motion (unless the temperature is at absolute zero). In a gas, the molecules move at constant speed in a straight line between collisions. Only a collision with another molecule causes a change of speed and direction. The collisions are perfectly elastic, i.e. there is no change in the total kinetic energy.

Gas Molecule Container

(a) Pressure and temperature

When the temperature is increased, the kinetic energy of the molecules increases. This means that the molecules move faster. As a result, the collisions with the walls are **harder** and **more frequent**. Each of these two effects increases the force on the walls and therefore the pressure increases, i.e. the law of pressures is explained.

(b) Pressure and volume

When the volume of the container is decreased, there is less space for the molecules to move around in. As a result, they collide with the walls more frequently (but **not** harder, as the temperature is constant). This increases the force on the walls and so the pressure increases, i.e. Boyle's Law is explained.

(c) Temperature and volume

When the temperature of the gas is increased, the molecules move around faster (greater E_k). As a result, each collision with the walls is harder. However, the force and pressure on the walls remains constant. There must therefore be fewer collisions per second. The only way of achieving this is for the molecules to move further apart – the volume increases i.e. Charles' Law is explained.

Quick Test 16

1. A gas occupies a volume of 2.5 m³ when it is in a room at a temperature of 25°C. Its temperature is now raised to 100°C. Calculate its new volume.
2. A sample of gas occupies a volume of 7.2 m³. The gas is at a pressure of 1.5×10^5 Pa and a temperature of 27°C. The sample is now heated to a temperature of 127°C and the pressure is increased to 2.5×10^5 Pa. Calculate the new volume.
3. Use the kinetic model to explain what happens to the pressure exerted by a gas as its temperature decreases.

Electric Charges and Fields

Types of electric charge

There are two types of electric charge – **positive** charges and **negative** charges. The tiny particles we call **electrons** (which orbit the nuclei of atoms) carry negative charge. The tiny particles we call **protons** (which, along with **neutrons**, form the nucleus of atoms) carry positive charge. Most objects have **equal** numbers of protons and electrons and so are electrically neutral.

A very large charge can build up on a Van de Graaf generator

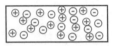

An object becomes **negatively** charged when it has **extra** electrons added to it.

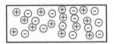 

An object becomes **positively** charged when it has some electrons **removed** from it.

 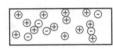

Opposite charges attract each other.   Like charges repel each other.

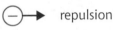

Opposite charges attract each other. Like charges repel each other.

$(+)\rightarrow \leftarrow(-)$ attraction $\leftarrow(+)$ $(+)\rightarrow$ repulsion

$\leftarrow(-)$ $(-)\rightarrow$ repulsion

Top Tip

Static electricity is used in many different ways. It is worth reading more about its use in precipitators, xerography, paint spraying, accelerators, ink-jet printing and electrostatic propulsion.

Charge, Q, is measured in **coulombs (C)** and is a **scalar** quantity. 6.25×10^{18} electrons are needed to make up one coulomb of charge. Each electron therefore carries only a tiny fraction of a coulomb, i.e. the charge carried by each electron is -1.6×10^{-19} C.

Electric fields

There is an electric field in a region of space if a **charge** experiences an electrostatic **force** when it is placed there.

An **electric field line** shows the direction of force acting on an **imaginary** positive charge placed in the electric field – the arrows on electric field lines point from positive to negative. The overall picture produced by drawing a number of these field lines is called an **electric field pattern**.

Examples of electric field patterns

Positive point charge

The imaginary positive test charge experiences a force 'outwards' from the central charge which is causing the electric field.

Negative point charge

The imaginary positive test charge experiences a force 'inwards' towards the central charge.

The previous two patterns are called **radial fields.** (Their lines are like the radii of a circle.)

Positive and negative point charges   Parallel charged plates

The field lines are equally spaced between the parallel plates.
This means the field strength is constant. This is called a **uniform field**.

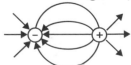

Moving a charged object in an electric field

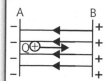

Here a positive charge, Q, is being pushed from the negative plate, A, on the left towards the positive plate, B, on the right. (It has to be **pushed** to move it to the right because the electric field is exerting a force on it towards plate A.)

When a positive charge is moved against the direction of the electric field's force, as in the diagram, energy is stored as electrical potential energy (like squashing a spring).

When the charge 'Q' is then released from plate B, its electrical potential energy, E_p, changes to kinetic energy, E_k, and the charge speeds up (accelerates) as it moves back towards plate A.

Definition of potential difference

The potential difference between two points is the work done in moving one coulomb of charge between the two points.

This means that if one **joule** of work is done moving one **coulomb** of charge between two points, the potential difference between the two points is one **volt**.

i.e. | 1 volt = 1 joule per coulomb   OR $1\ V = 1\ J\ C^{-1}$

This gives the following relationship:

| W = Q V |

W = 'work done' in joules (J) i.e. the energy transferred.
Q = charge in coulombs (C)
V = potential difference in volts (V)

Electron Gun

Example

In an electron gun, an electron is accelerated from rest through a potential difference of 2 000 V.

Calculate   (a) the kinetic energy, E_k, gained,
              (b) the final speed of the electron.

(Data: mass of an electron = 9.11×10^{-31} kg; size of charge on an electron = 1.6×10^{-19} C)

Answer

(a) $E_k = \frac{1}{2} mv^2 = QV = 1.6 \times 10^{-19} \times 2000 = 3.2 \times 10^{-16}$ J

(b) $\frac{1}{2} mv^2 = 3.2 \times 10^{-16}$ J

$\Rightarrow v^2 = 2 \times 3.2 \times 10^{-16}/9.11 \times 10^{-31}$

$\Rightarrow v = 2.65 \times 10^7$ ms^{-1}

Top Tip

Try working out the effect of doubling the potential difference to 4 000 V. You should find that the speed of the electrons increases but does not double.

Quick Test 17

1. A balloon is rubbed on a jersey and becomes positively charged. Explain what has happened to cause this.
2. (a) What symbol is used for electric charge?
   (b) What are the units of charge?
   (c) Is charge a scalar quantity or a vector quantity?
3. How many electrons are needed to produce a total charge of 4.8×10^{-15} C?
4. A charge of 7.4×10^{-14} C is moved through a potential difference of 60 V. How much work is done?
5. An electron is accelerated from rest through a potential difference of 400 V.
   Calculate the final speed of the electron.

Circuits: Series and Parallel Connections

When an electric field is applied to a conductor (for example by connecting a battery across its ends), the force on the free electric charges (electrons) in the conductor causes them to move.

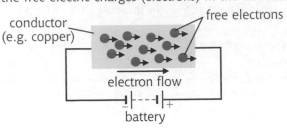

conductor (e.g. copper)

free electrons

electron flow

battery

The flow of charge in the conductor is called the **electric current**. Current is measured in amperes (A). One ampere means that one coulomb of charge passes every second, i.e.

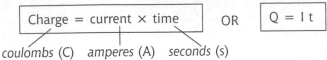

Charge = current × time   OR $Q = I t$

coulombs (C)   *amperes* (A)   *seconds* (s)

Some basic facts about circuits

- A series circuit is one which has only one path for the electrons to flow through.
- The current is the same everywhere in a series circuit.
- In a series circuit the current stops everywhere if the circuit is broken anywhere.
- Parallel connections are where there is a split in the wiring, giving the electrons a 'choice' of paths round the circuit.
- Circuits are temporarily broken and ammeters are connected in series to measure current.
- Voltmeters are connected in parallel to measure potential difference.

Connecting resistors

When **resistors** are connected together in **series** (i.e. only one path for the current), the total resistance is found by **adding** the individual resistances.

picture

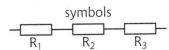

symbols

R_1 R_2 R_3

The series formula is   Total resistance, $R_T = R_1 + R_2 + R_3 + \ldots\ldots\ldots$

The supply voltage V_s, **divides** up across the resistors **in proportion** to their values of **resistance**.

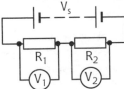

V_s

R_1 R_2

V_1 V_2

$$V_1 = \frac{R_1}{(R_1 + R_2)} \times V_s$$ and $$V_2 = \frac{R_2}{(R_1 + R_2)} \times V_s$$

This is often referred to as a **potential divider** circuit.

When resistors are connected together in **parallel** (i.e. more than one path for the current), the total resistance is found using the following relationship:

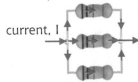

current, I

R_1

R_2

R_3

$$\frac{1}{R_T} = \frac{1}{R_1} + \frac{1}{R_2} + \frac{1}{R_3} + \ldots$$

Top Tip

You must remember that the left-hand side of this formula is **one over** the total resistance.

In some questions you will need to use both the series and parallel formulas.

Example

Three resistors are connected together as shown.
What is the total resistance between points X and Y?

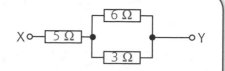

Answer

The 6 Ω and 3 Ω resistors are in parallel, producing a combined parallel resistance of 2 Ω.
This 2 Ω is in series with the 5 Ω resistor – giving an overall total of 7 Ω.

Ohm's Law

Ohm's law is the relationship between the **current**, I, in a resistor and the
potential difference, V, across the resistor.

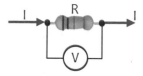

For a fixed resistor, the current and the potential difference are **directly proportional** to
each other. The constant of proportionality is called the **resistance, R**, of the resistor and
is measured in ohms (Ω).

Potential difference = current × resistance

volts (V)   amperes (A)   ohms (Ω)

OR $V = I\,R$

For a **fixed** resistor the voltage/current graph is a straight diagonal line through the origin and the gradient
equals the value of the resistance.

Example

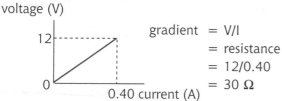

voltage (V)

gradient = V/I
= resistance
= 12/0.40
= 30 Ω

However, for some components,
such as filament lamps, the value
of resistance changes as the
current changes.
The graph is a curve.

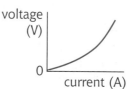

Power in electricity

Power = current × potential difference

watts (W)   amperes (A)   volts (V)

OR $P = I\,V$

It can also be shown that

Power = current² × resistance

watts (W)   amperes (A)   ohms (Ω)

OR $P = I^2\,R$

and that

$$\text{Power} = \frac{\text{voltage}^2}{\text{resistance}}$$

volts (V)
ohms (Ω)
watts (W)

OR $P = \dfrac{V^2}{R}$

Top Tip

Using P = IV and
V = IR, try to show
that these two
formulas are correct.

1. What is the total resistance of each of the following combinations of resistors?

(a) 7.6 Ω 5.2 Ω 11.3 Ω

(b)
12 Ω
4.0 Ω

(c)
6.0 kΩ   5.0 kΩ   12 kΩ
3.0 kΩ   4.0 kΩ

2. Calculate the values of V_1 and V_2 in this circuit.

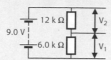

9.0 V   12 kΩ V_2
        6.0 kΩ V_1

3. A wire has a current in it of 3.6 A when connected
   to a potential difference of 12 V. Calculate
   (a) the resistance of the wire
   (b) the power output.

e.m.f. and Internal Resistance

e.m.f

The e.m.f. of any electrical supply means, 'the number of joules of electrical energy given to each coulomb of electric charge as it passes through the supply.'
The units of e.m.f. are **joules per coulomb** (J C^{-1}) or Volts (V).

Example
A 12 V car battery provides 12 J of energy to every coulomb of charge passing through it, whereas a 1.5 V cell gives each coulomb only 1.5 J of energy. This is why the car battery produces a greater current than the cell when they are connected to identical circuits.

Car battery

Internal resistance, r

All power supplies act as if they have a resistance inside them connected in series with the e.m.f. – this resistance is called the **internal resistance** and is measured in ohms.
You will find it useful to keep this picture in your mind for any power supply.

positive terminal   negative terminal

Formula
When a supply of e.m.f., E, and internal resistance, r, is connected to a circuit of resistance, R, it causes a current, I, throughout the circuit.

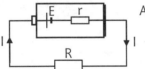

Applying Ohm's law to this circuit gives:

$$\text{voltage} = \text{current} \times \text{resistance}$$
$$\text{e.m.f.} = I(R + r)$$
$$E = IR + Ir \quad \text{(the total resistance of the series circuit is } \{R + r\})$$
$$\text{OR} \quad \boxed{E = V + Ir}$$

Terminal voltage – the voltage across the terminals of the power supply.

'Lost volts' – the voltage dropped across the internal resistance, r.

> **Top Tip**
> Think of this as a potential divider circuit with 'r' and 'R' as the two resistors in series.

Experiment
The circuit shown below can be used to find the e.m.f. and internal resistence of a cell.

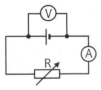

The value of the variable resistor, R, is altered and a series of readings of the terminal voltage, V, and current, I, is recorded. The results produce this graph.
The e.m.f. of the supply is the **intercept on the voltage axis**, i.e. the output voltage when there is no current in the circuit. The internal resistance of the supply is equal to the **negative of the gradient** of this graph.

Output Voltage, V /volts — the e.m.f.

gradient = – internal resistance

Current, I/A

Short circuit current

To 'short circuit' a power supply means to connect a zero resistance path between its terminals. The resistance in the circuit is then solely due to the internal resistance.

The current is then calculated from:

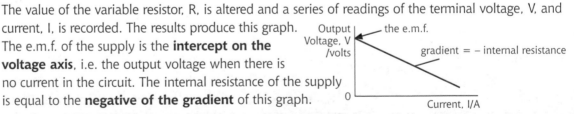

$$I_{\text{short circuit}} = \frac{\text{e.m.f.}}{\text{internal resistance}} = \frac{E}{r}$$

Internal resistance and power output

When a power supply is connected to a load (i.e. the resistance of the external circuit), current causes the internal resistance to heat up as well as causing useful power output in the load. This means there is **wasted** power output within the power supply.

Power supply

It is essential to understand that the internal resistance needs to be included in order to calculate the current in the circuit, but that only the value of the load resistance is used to calculate the useful output power.

Load

Calculations can be carried out to show that:

- the greater the load resistance, the greater the proportion of the e.m.f. across the load;
- maximum power is transferred to the load when load resistance = internal resistance. However, this does not mean maximum efficiency, since as much power is dissipated as heat inside the power supply as is produced as useful power output from the load.

Example

Two cells each of e.m.f. 1.5 V and internal resistance 1.2 Ω are connected in series with each other and a load resistor of value 3.6 Ω. Calculate:

(a) the current in the circuit;
(b) the potential difference across the load resistor;
(c) the power output from the load resistor;
(d) the power wasted within each cell.

The 3.6 Ω load resistor is now replaced by one of value 2.4 Ω. Calculate:
(e) the potential difference across the new load resistor;
(f) the power output from the new load resistor.

> **Top Tip**
>
> Draw sketches of the whole circuit including each cell, showing both its e.m.f. and its internal resistance.

Answer

(a)
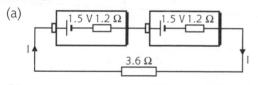

Total e.m.f. = 1.5 V + 1.5 V = 3.0 V
Total circuit resistance = 1.2 Ω + 1.2 Ω + 3.6 Ω
$\qquad\qquad$ = 6.0 Ω

Using Ohm's Law:
$\qquad I = V/R = 3.0/6.0 = 0.50$ A

(b) $V_{load} = I\,R = 0.50 \times 3.6 = 1.8$ V

(c) $P_{load} = I^2 R = 0.5^2 \times 3.6 = 0.90$ W $\qquad$ OR $\qquad P_{load} = I\,V = 0.50 \times 1.8 = 0.90$ W

(d) $P_r = I^2 R = 0.5^2 \times 1.2 = 0.30$ W

(e) $I_2 = V/R = 3.0/4.8 = 0.625$ A
$\qquad\qquad V_{load} = I\,R = 0.625 \times 2.4 = 1.5$ V

(f) $P_{load} = I^2 R = 0.625^2 \times 2.4 = 0.9375 = 0.94$ W
$\qquad$ OR $\qquad P_{load} = I\,V = 0.625 \times 1.5 = 0.94$ W

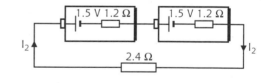

Quick Test 19

1. A power supply has an e.m.f. of 9.0 V. What is meant by 'an e.m.f. of 9.0 V'?
2. The e.m.f. of a cell is 1.5 V. The voltage across its terminals is 1.4 V when the current is 0.20 A.
   Calculate (a) the lost volts;
   $\qquad\qquad$ (b) the internal resistance.
3. Two cells each of e.m.f. 1.6 V and internal resistance 0.50 Ω are connected in series with each other and a load resistor of value 15 Ω.
   Calculate (a) the current in the circuit;
   $\qquad\qquad$ (b) the potential difference across the load resistor;
   $\qquad\qquad$ (c) the power output from the load resistor.

The Wheatstone Bridge

A Wheatstone bridge circuit

A Wheatstone bridge circuit consists of four resistors connected as shown.

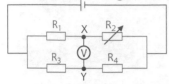

A Wheatstone bridge is two potential dividers connected in parallel with each other.

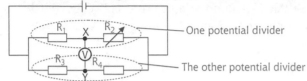

— One potential divider

— The other potential divider

One potential divider is made up of R_1 and R_2 connected in series with each other.
The other potential divider is made up of R_3 and R_4 connected in series with each other.
The bridge is said to be **balanced** when the resistors in the two potential dividers are in the **same ratio** as each other, i.e.

$$\frac{R_1}{R_2} = \frac{R_3}{R_4}$$

A sensitive voltmeter is used to detect when the bridge is balanced. It is connected between the mid-points, X and Y, of the two potential dividers.
When the bridge is balanced, the voltmeter reads **zero**.

Top Tip

Use pages 38–39 to remind yourself of how to find values of p.d. in potential divider circuits.

Find the value of an unknown resistance

One use of a Wheatstone bridge circuit is to find the value of an **unknown** resistance.

Procedure
- The circuit is set up with three resistors of known values – one needs to be a variable resistor.
- The unknown resistance is connected into the circuit as the fourth resistor.
- The variable resistor is adjusted until the voltmeter reads zero, i.e. the bridge is balanced.
- The four resistors are then in the ratio of $\dfrac{R_1}{R_2} = \dfrac{R_3}{R_4}$

- Substituting the three known values then allows the unknown resistance to be calculated.

Note
- The balance condition does not depend on the value of the supply voltage.
- The meter must be sensitive to detect zero precisely.
- The circuit is sometimes drawn in a diamond shape or rotated through 90°.

Example

Thermistors

A thermistor is connected in a Wheatstone bridge circuit as shown.
At a certain temperature, the voltmeter reads zero when the variable resistor is set to a value of 8.4 kΩ. Calculate the value of the resistance, R_t, of the thermistor at this temperature.

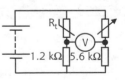

Answer
At balance, $R_1/R_2 = R_3/R_4$
$\Rightarrow$ $1.2/R_t = 5.6/8.4$
$\Rightarrow$ $R_t = (1.2 \times 8.4)/5.6$
$\Rightarrow$ $R_t = 1.8$ kΩ

The Wheatstone bridge slightly out-of-balance

When one of the resistors is varied **slightly** above and below its balance value, the voltmeter shows a reading which is **directly proportional** to the **change** in resistance. Here is the graph that shows this.

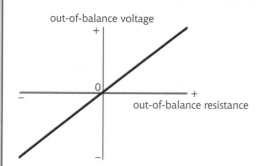

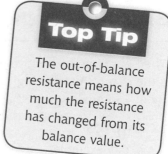

Top Tip

The out-of-balance resistance means how much the resistance has changed from its balance value.

This direct proportionality (or linear) relationship allows the Wheatstone bridge to be used as a type of measuring device when operated **near to its balance point**.

Examples

1. One of the resistors can be replaced by a **thermistor**. The bridge then acts as a **thermometer** giving higher (positive or negative) voltages for greater temperature changes.

2. One of the resistors can be replaced by a **light dependent resistor** (**LDR**). The bridge then acts as a **light meter** giving higher (positive or negative) output voltages for greater changes in light level.

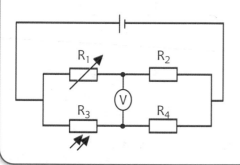

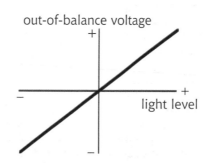

An LDR

Quick Test 20

1. What is the minimum number of resistors needed to build a Wheatstone bridge circuit?

2. Say whether each of the following statements about the Wheatstone bridge is true or false.
   (a) A Wheatstone bridge is made of two potential dividers connected in series.
   (b) At balance, the p.d. between the two mid points of the potential dividers is zero.
   (c) To detect the balance point precisely, the voltmeter should be very sensitive.
   (d) Increasing the supply voltage changes the balance condition.
   (e) A Wheatstone bridge can be used slightly out-of-balance to measure small changes in light level or temperature.

3. An LDR is connected in a Wheatstone bridge circuit as shown.
   At a certain light level, the voltmeter reads zero when the variable resistor is set to a value of 22 kΩ.
   Calculate the value of the LDR's resistance, R_{LDR}, at this temperature.

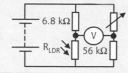

Alternating Current and Voltage

Examples of oscilloscope traces for d.c. and a.c.

When electrons keep moving round a circuit in the **same** direction, we say that there is **direct current** (**d.c.**) in the circuit. **Alternating current** (a.c.) is when electrons **regularly reverse** their direction of flow round a circuit – in other words, they oscillate in the wires of the circuit.

Examples

d.c. examples

0 V ·······························

A fixed voltage
(e.g. from a cell)
with the oscilloscope's
timebase switched off.

0 V ·······························

A fixed voltage
with the oscilloscope's
timebase switched on.

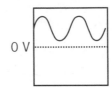

0 V ·······························

Unsmoothed, fluctuating d.c.
(this is not a.c. because the
electrons do not reverse their
direction of flow as the trace never
falls below the 0 V level).

a.c. examples

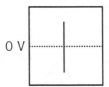

0 V ·······························

Any shape of wave
timebase off.

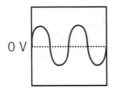

0 V ·······························

Sine wave
timebase on.

0 V ·······························

Triangular/sawtooth
timebase on.

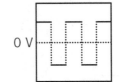

0 V ·······························

Square wave
timebase on.

Higher Physics deals mainly with the **sine wave** – for example, the mains voltage is a sine wave.

Measuring peak voltage of an a.c. signal using an oscilloscope

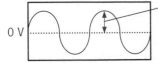

0 V ·······························

peak voltage, V_p

The peak voltage is the **amplitude** of the wave, i.e. the value between the zero level and the top of the wave.

1. On the screen of the oscilloscope, measure the number of divisions from the zero line to the peak.
2. Multiply this number of divisions by the y-gain setting on the oscilloscope.

Example

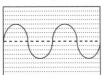

The Y-gain setting is 0.5 V cm^{-1},
no. of divisions to peak = 4
peak voltage = 4 × 0.5
= 2.0 volts

Measuring the frequency of an a.c. signal

1. As accurately as possible, find the number of divisions **across** the screen for **one** wave.
2. Multiply the answer to part 1 by the timebase setting to get the time for one wave – this is the **period** of the wave.
3. Calculate the frequency from

$$\text{frequency} = \frac{1}{\text{period}}$$

hertz ⎤ ⎣ *seconds*

Example

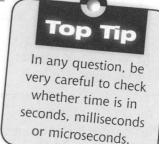

Time base setting = 0.5 ms per div,
1. there are 5.0 divisions for each wave
2. time for 1 wave = $5.0 \times 0.5 \times 10^{-3}$
$$= 2.5 \times 10^{-3}\,\text{s}$$
3. frequency = $\dfrac{1}{2.5 \times 10^{-3}} = 400\ \text{Hz}$

Top Tip

In any question, be very careful to check whether time is in seconds, milliseconds or microseconds.

Peak and r.m.s. values

The instantaneous value of any alternating voltage is constantly increasing up and down between zero and its maximum, the **peak** value, V_p. The constant (or effective) voltage that has the **same** overall effect (for example, in heating a wire) has a value **less than** the peak voltage. This effective value of an alternating voltage is called the **r.m.s.** (root mean square) voltage or $V_{r.m.s.}$.

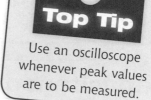

$V_{r.m.s}$ is less than V_p

Similarly, the current in the circuit has a peak value, I_p, and an effective value, $I_{r.m.s.}$. Voltmeters and ammeters measure r.m.s. values.

To calculate **average** power, r.m.s. values must be used, i.e. $\boxed{\text{Power} = V_{r.m.s} \times I_{r.m.s}}$

For a sine wave, the relationships between r.m.s. values and peak values are: $V_{rms} = \dfrac{V_p}{\sqrt{2}}$ and $I_{rms} = \dfrac{I_p}{\sqrt{2}}$

Top Tip

Use an oscilloscope whenever peak values are to be measured.

Relationship between current and frequency for a resistor

- A resistor is connected to a signal generator as shown.
- The output voltage is kept constant (its value is monitored using the oscilloscope).
- The frequency of the signal is varied and the current measured.

Results:

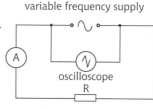

The value of the current **stays the same** at all frequencies for a resistor.

1. Say whether each of the following statements is true or false.
   (a) d.c. means that electrons do not change their direction of flow round a circuit.
   (b) a.c. means that electrons can flow both ways round a circuit.
   (c) The peak voltage of a sine wave is always greater than its r.m.s. voltage.
   (d) A voltmeter can be used to measure peak voltage directly.
2. On an oscilloscope screen, the peak of a signal is 4 divisions above the zero level. The y-gain is set to 5.0 V/div. Calculate the peak voltage.
3. The width of a wave on the screen of an oscilloscope is 2.0 div. The timebase is set at 4.0 ms/div. Calculate the frequency of the signal.
4. For British mains, $V_{r.m.s.} = 230$ V. Calculate the peak voltage.

Capacitors

What is a capacitor?

A capacitor is an electrical component which can **store electric charge**. It is often constructed in the form of two parallel plates, separated by an insulating material such as air, paper or plastic. Its circuit symbol is

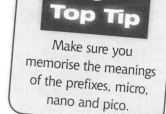

Capacitors

When a capacitor is connected to a power supply, the potential difference of the supply causes electrons to be **repelled** from its **negative** terminal on to one of the plates of the capacitor. At the same time, electrons are **attracted** off the other plate of the capacitor towards the **positive** terminal of the supply, i.e. **work** is done by the supply in order to store charge in the capacitor.

The charging process in pictures

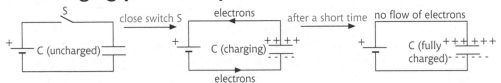

After a short time, there is no longer a flow of electrons in the circuit. This is because the 'push' of the supply voltage is balanced by an equal and opposite force on the electrons due to the potential difference which has built up across the plates of the capacitor. (Note that it is incorrect to say that the capacitor is 'full', because more charge could be forced on to its plates by increasing the supply voltage.)

Capacitance formula

The value of the capacitance of a capacitor means the **number of coulombs** of charge it stores for every **volt** across its plates, i.e.

$$\text{Capacitance} = \frac{\text{Charge}}{\text{Voltage}} \quad \text{or} \quad C = \frac{Q}{V}$$

The units of capacitance are coulombs per volt (C V^{-1}) or **farads** (F).
In practice, one farad is large and most capacitors are only fractions of a farad.

1 microfarad	=	1.0 μF	=	1.0×10^{-6} F
1 nanofarad	=	1.0 nF	=	1.0×10^{-9} F
1 picofarad	=	1.0 pF	=	1.0×10^{-12} F

Top Tip

Make sure you memorise the meanings of the prefixes, micro, nano and pico.

More about charging and discharging

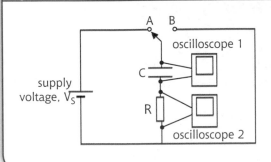

When the switch is connected to position A, the capacitor charges from the supply through resistor R.
When the switch is then moved to position B, the capacitor discharges through resistor R.
Oscilloscope 1 displays a voltage/time graph for the capacitor.
Oscilloscope 2 displays a current/time graph for the circuit.
(Although an oscilloscope always displays a graph of voltage against time, the voltage across a fixed resistor is proportional to the current through it (because I = V/R).)

Graphs for a capacitor charging

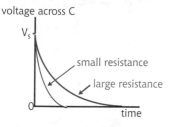

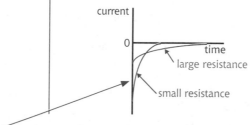

Graphs for a capacitor discharging

The **discharge current** is shown as **negative** because the electrons flow the **opposite** way round the circuit during discharge. Although electric charge flows through the cell, through the connecting wires and through the resistor during charging and discharging, it **cannot flow** through the capacitor since there is an insulator between the plates. Charge can only flow on to and off the plates of the capacitor.

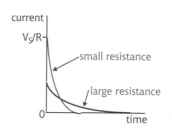
Charging a capacitor with a constant current

Normally, current decreases during charging. A **constant** charging current can be achieved by either:

(a) using a special type of power supply

or

(b) steadily decreasing the resistance of a variable resistor in series with the capacitor.

When current, I, is constant, the charge, Q, stored on the capacitor is directly proportional to time, t.

Experimental results produce this graph:

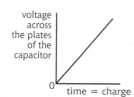

The graph is a straight line through the origin, which proves that the charge on a capacitor is directly proportional to the p.d. across its plates.

This result matches the formula on page 46, which says that the ratio of charge/voltage is constant and is equal to the capacitance of the capacitor. (= 1/gradient of this graph).

Quick Test 22

1. A capacitor stores a charge of 4.95×10^{-5} C when connected to a cell of e.m.f. 1.5 V. Calculate its capacitance.
2. A 47 µF capacitor is connected to a 9.0 V battery. Calculate the charge stored on the capacitor.
3. An uncharged capacitor of capacitance 220 pF is connected in series with a 6.0 V battery and a 120 Ω resistor. Calculate the initial charging current.

Capacitors in Circuits

Energy stored in a capacitor

As a result of storing charge on its plates, a **charged** capacitor also **stores** electrical **energy**. This energy can be released to do useful work – for example, in the electronic flash of a camera.

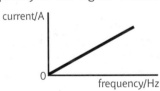

The **area** under a voltage/charge graph gives the energy stored in a capacitor. This produces one of three formulas for the energy stored in a capacitor.

$$E = \tfrac{1}{2} \text{ charge} \times \text{voltage} \quad \text{OR} \quad \boxed{E = \tfrac{1}{2} Q V}$$

Substituting first 'Q = C V' and then 'V = Q/C' gives the other two formulas for the energy stored in a capacitor.

$$\boxed{E = \tfrac{1}{2} C V^2} \quad \text{and} \quad \boxed{E = \tfrac{1}{2} \frac{Q^2}{C}}$$

Top Tip

Do not confuse this with the formula W = Q V given on page 37.

Capacitors and a.c.

When a capacitor is connected to a d.c. supply there is only a **brief**, **charging** current. The flow of charge cannot continue because the circuit is broken between the plates of the capacitor. However, with an a.c. supply, the capacitor charges up one way, then discharges and charges up the opposite way. This process is **continuously repeated** – alternating current is 'permitted'.

Relationship between current and frequency for a capacitor

- A capacitor is connected in a circuit as shown.
- The output voltage is kept constant (its value is monitored using the oscilloscope).
- The frequency of the signal is varied and the current measured.

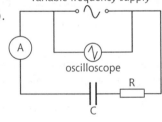

Results

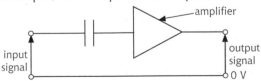

Current is **directly proportional** to frequency for a capacitor.

More uses of capacitors

1. As a block to d.c.

The **insulator** between the plates of a capacitor **prevents** direct current. When a signal is a combination of d.c. and a.c. only the a.c. part will be allowed to pass when a capacitor is connected in series. This can be used, for example, at the input of an amplifier:

input signal — ‖ — amplifier — output signal — 0 V

Another example of this use is at the input stage of an oscilloscope where there is an a.c./d.c. switch which allows the selection of only the a.c. component of a signal.

2. As a filter

As a filter to allow only high frequencies to pass.

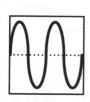

The capacitor 'allows' a **greater current at higher frequencies** than at lower frequencies. As a result, this loudspeaker mainly emits high frequency notes (and so it is often referred to as a 'tweeter').

3. To smooth pulsed d.c.

Changing a.c. into d.c. is called **rectification**. This process produces pulsed or rippled d.c.
A capacitor, connected in parallel, can smooth out these pulses by repeatedly charging (when the diode is conducting) and discharging (when the diode is not conducting).

Example

Half-wave rectification (using a single diode):

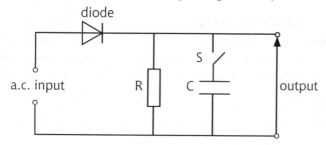

An oscilloscope displays the following traces:

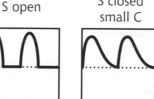

input signal

output signal

| S open | S closed small C | S closed large C |

unsmoothed pulses of d.c.

slightly smoothed pulses of d.c.

smooth d.c.

Top Tip

Try to memorise at least one practical use of a capacitor.

A larger capacitance allows more charge to be stored on the capacitor and produces better smoothing of the d.c. pulses.

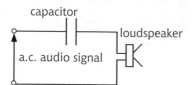

Quick Test 23

1. A capacitor stores 6.4×10^{-4} C of charge when connected to a battery of e.m.f. 6.0 V.
   Calculate the energy stored in the capacitor.
2. A capacitor of capacitance 430 nF is connected to a 12 V battery.
   Calculate the energy stored in the capacitor.
3. A capacitor of capacitance 6800 μF is storing 3.4×10^{-4} C of charge on its plates.
   Calculate the energy stored in this capacitor.
4. A capacitor is connected in series with a resistor and an a.c. supply. The frequency of the supply is gradually increased.
   Describe what happens to the current in the circuit.
5. Give a brief description of a use for a capacitor.

The Operational Amplifier (op-amp) 1

The op-amp

An operational amplifier (or **op-amp**) can change the size of an input voltage (i.e. it can **amplify voltage**). The symbol for an op-amp is shown in blue,

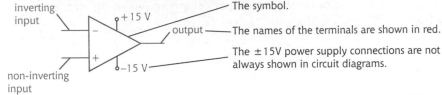

inverting input

+15 V

The symbol.

output

The names of the terminals are shown in red.

−15 V

The ±15V power supply connections are not always shown in circuit diagrams.

non-inverting input

An op-amp 'chip'

To use an op-amp, resistors are connected to its terminals to make **amplifier circuits**.
In Higher Physics there are **two** types of op-amp circuits (or **modes**) you need to learn.

The op-amp in inverting mode

1. Connect a resistor between the output and the inverting input – the feedback resistor, R_f.
2. Connect another resistor at the inverting input – the input resistor, R_1.
3. Connect the non-inverting input directly to the zero volts line (the earth/ground line).

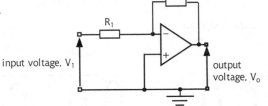

input voltage, V_1

output voltage, V_o

Experiments show that:

1. the output voltage is always the **opposite** sign to the input voltage (the signal is **inverted**), i.e.
   a positive input voltage produces a negative output voltage, and
   a negative input voltage produces a positive output voltage;
2. the relationship between the two voltage values and the two resistor values is:

$$\frac{V_o}{V_1} = -\frac{R_f}{R_1}$$

   This is often called the inverting mode gain formula
3. the output voltage cannot exceed the supply voltage.

In practice, the maximum output voltage is approximately two volts less than the supply voltage. Using a normal ±15 V supply, this means that the maximum output voltage is approximately ±13 V. We say that the **amplifier saturates** at this maximum output voltage. (Note that it is wrong Physics to say that the voltage saturates.)

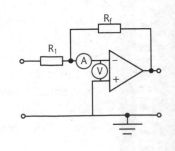

Top Tip

Make sure you can:
- recognise this mode,
- draw the circuit from scratch.

The ideal op-amp

An **ideal** op-amp is defined as one where:
- the input current is **zero**, i.e. it has **infinite** input resistance;
- there is **zero potential difference** between the inverting and non-inverting **input terminals**, i.e. both input pins are at the same potential.

R_f

R_1

These statements mean that, for an ideal op-amp, both the ammeter and the voltmeter in this circuit read zero.

Note that the terminals (or pins) are the ones at the op-amp 'chip' and not the input terminals of the amplifier circuit.

Alternating voltage applied to an op-amp in inverting mode

An alternating input signal produces an alternating output signal of the **same frequency** but, because the op-amp inverts the signal, the output a.c. waveform is **180° out of phase** with the input waveform – this means that a positive peak input voltage produces a negative peak output voltage (and vice versa). The inverting mode gain formula is still obeyed and so the peak voltages are in the ratio of R_f/R_1.

Example

A sine wave signal is connected to the input terminals of this op-amp circuit.

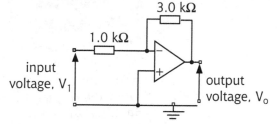

Voltage gain $= -R_f/R_1$
$= -3.0/1.0$
$= -3$

A double-beam oscilloscope displays both the input and output signals as follows.

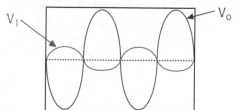

Top Tip

Remember:
- invertion,
- amplification,
- frequency constant.

Saturation/square waves

When the gain is high enough, the amplifier saturates when the output voltage is about ±13 V. A **high gain** therefore produces a **square wave output** voltage, i.e.

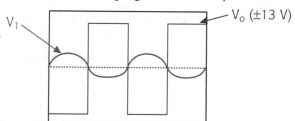

V_o (±13 V)

Quick Test 24

1. An op-amp is connected in inverting mode. The following values are known:
   $R_1 = 1.5 \text{ k}\Omega$ $R_f = 60 \text{ k}\Omega$ $V_1 = 0.25 \text{ V}$
   Calculate the output voltage.
2. An op-amp is connected in inverting mode. The following values are known:
   $R_1 = 1.5 \text{ k}\Omega$ $R_f = 120 \text{ k}\Omega$ $V_1 = -0.25 \text{ V}$
   Calculate the output voltage.
3. An op-amp is connected in inverting mode. The following values are known:
   $R_1 = 6.8 \text{ k}\Omega$ $R_f = 2.5 \text{ M}\Omega$ $V_o = 6.5 \text{ V}$
   Calculate the input voltage.
4. A sine wave of peak value 5.0 V is connected to the input terminals of an op-amp in inverting mode. The gain of the op-amp is – 4. Is the output a square wave? Explain your answer.

The Operational Amplifier (op-amp) 2

The op-amp in differential mode

The circuit diagram for this mode is

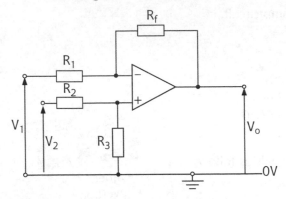

In Higher Physics we only use this circuit when,

$$\frac{R_f}{R_1} = \frac{R_3}{R_2}$$

Experiments show that this circuit:
- finds the difference between V_2 and V_1,
- and multiplies this difference by the gain factor R_f/R_1.

i.e. in the differential mode, an op-amp amplifies the potential difference between its two inputs.

Formula

$$V_0 = (V_2 - V_1) \times \frac{R_f}{R_1}$$

This means that when **V_2 is greater than V_1**, the output voltage is **positive** and, when V_2 is less than V_1, the output voltage is negative.

The op-amp can **saturate** (giving an output voltage of approximately $\pm 13V$) if either there is a **very high gain** or if the **difference** between the input voltages is large.

Example

An op-amp is connected in a circuit as shown.

Calculate the output voltage, V_0.

Answer

$V_1 = 6.5\ V \qquad V_2 = 6.3\ V \qquad$ gain $= R_f/R_1 = R_3/R_2 = 10/2 = 5$

$V_0 = (V_2 - V_1) \times R_f/R_1$

$\quad = (6.3 - 6.5) \times 5$

$\quad = -1.0\ V$

Automatic monitoring and control

An electronic circuit can measure, for example, the surrounding temperature and can then produce an appropriate output to change the situation.

Example

A circuit can monitor the temperature in a greenhouse. It can then also automatically open and close a window as necessary to help keep the temperature constant. The basic design of the circuit is shown below.

How the circuit operates

When the Wheatstone bridge is balanced (by adjusting the variable resistor) at the desired temperature, the potential difference input to the op-amp (in differential mode) is zero. The output is therefore also zero, neither transistor is switched on and the motor is stopped.

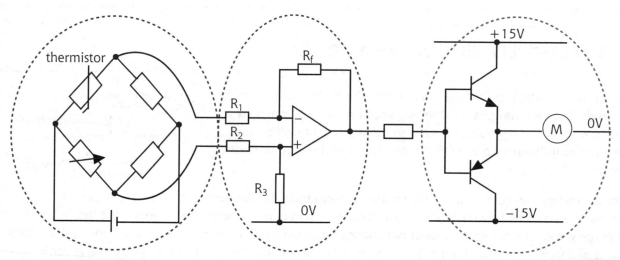

When the temperature rises, the resistance of the thermistor **decreases**; this puts the Wheatstone bridge **out-of-balance**. The out-of-balance voltage is then **amplified** by the op-amp giving a **high enough** output voltage to switch on one of the transistors. As a result the motor is switched on and moves the window open.

When the temperature falls, the out-of-balance voltage has the **opposite** sign. This causes the other transistor to be switched on, operating the motor in the opposite direction, closing the window.

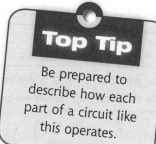

Top Tip

Be prepared to describe how each part of a circuit like this operates.

1. Calculate the output voltage, V_o.

2. Calculate the value of V_1.

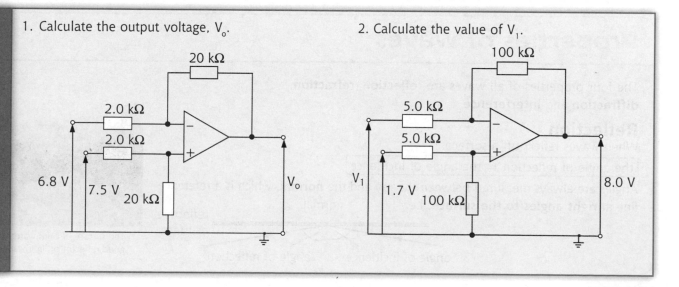

Wave Terms and Properties

Different kinds of waves

There are different kinds of wave, for example: water waves, sound waves, light waves, radio waves and other electromagnetic waves.

All **waves transfer energy** from place to place.

All waves have common characteristics (or properties) and terms describing them.

Terms used to describe waves

The **crest** of a wave is the top (or peak) of a wave. The bottom of a wave is called a **trough**. The **wavelength**, λ, is the distance (in metres) from a crest to the next crest (or from any part of a wave to the equivalent part on the next wave). The **amplitude** of a wave is the distance from the mid-position to the crest or to the trough.

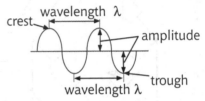

The **greater** the **energy** carried by a wave, the **greater** its **amplitude**. The **frequency**, **f**, of a wave means the **number** of waves each second. Frequency is a scalar quantity and its units are **hertz (Hz)**. Frequency is determined by the source of the waves and so frequency **does not change** when the waves move into a different medium. The speed (or velocity, v) of a wave is how fast (in m s⁻¹) the energy travels away from the source of the waves. The velocity is determined by the substance (or medium) the waves are travelling in. Once frequency and velocity have been determined by the source and the medium respectively, the wavelength can be calculated from the **wave formula**:

$$\text{wave velocity} = \text{frequency} \times \text{wavelength} \qquad \text{OR} \qquad v = f\,\lambda$$

The **period**, **T**, of a wave is the **time** for one **complete wave**.
The relationship between period and frequency is:

$$\text{period} = \frac{1}{\text{frequency}} \quad \text{OR} \quad T = \frac{1}{f} \quad \text{and} \quad \text{frequency} = \frac{1}{\text{period}} \quad \text{OR} \quad f = \frac{1}{T}$$

The units of period are seconds (s) when the units of frequency are hertz (Hz).

> **Top Tip**
>
> The amplitude is **half** the total height of the wave.

Properties of waves

The four properties of all waves are **reflection**, **refraction**, **diffraction** and **interference**.

Reflection
When waves reflect off a surface,

$$\boxed{\text{the angle of reflection} = \text{the angle of incidence}}$$

Angles are always measured between the ray and the **normal**, which is a reference line at **right angles to the surface**, i.e.

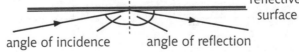

The surface of calm water is a good reflector of light waves.

Refraction

When a ray of light travels from one medium into another there is a **change** in its **speed** and (usually) its **direction** – this is called the refraction of light. Examples include a ray of light passing through a rectangular glass block:

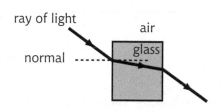

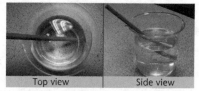

Refraction of light causes this pencil to appear to be 'broken'.

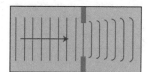
Diffraction

When waves meet an edge (or a gap), they bend round behind the edge. This property of waves is called **diffraction**. Waves bend round (diffract) **more** when their wavelength is **longer** (or when a gap is narrower), i.e.

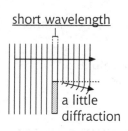

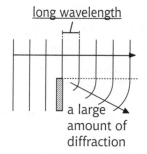

A wide gap causes a little diffraction.

A narrow gap causes a lot of diffraction.

Interference

When two (or more) waves meet each other they temporarily overlap, causing larger or smaller amplitudes, for example:

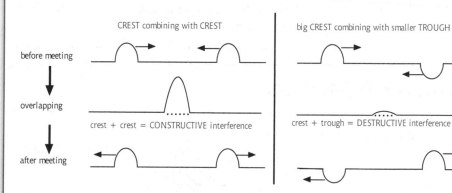

Although all types of waves can be reflected, refracted and diffracted, we can only prove that energy is travelling in waves by creating an interference pattern.

Quick Test 26

1. What changes about a wave when it carries more energy?

2. Calculate the wavelength of these waves.
3. Some sound waves have a period of 25 ms.
   (a) What does a period of 25 ms mean?
   (b) Calculate the frequency of the sound.
4. Which property proves that the energy is travelling as a wave?

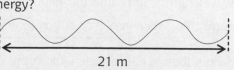

21 m

Phase, Path Difference and Interference

Waves in phase and out of phase

In phase

Waves are said to be **in phase** when a crest of one wave combines with the crest of another wave (and a trough from one wave combines with the trough of another wave), i.e.

Waves combining in phase.

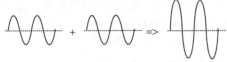

A wave of greater amplitude is produced.

Waves combining in phase produce constructive interference.

Out of phase

Waves are said to be perfectly **out of phase** when the crest of one wave combines with the trough of another wave, i.e.

Waves combining out of phase.

A wave of smaller amplitude is produced.

Waves combining out of phase produce destructive interference.

Coherence

Sources of waves are said to be **coherent** when they have the **same frequency** and are **in phase** with each other, for example, two loudspeakers connected (identically) in parallel to one signal generator.

Interference and path difference

Think of two coherent sources (S_1 and S_2) producing waves which overlap and produce an **interference pattern**:

Whether constructive interference or destructive interference occurs at any point in the pattern depends on the wavelength and the **different distances** to that point from the two sources.

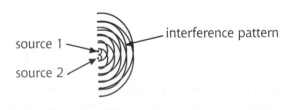

source 1
source 2
interference pattern

Here, at P, the waves combine in phase and so P is a point of constructive interference.

Here, at Q, the waves combine out of phase and so Q is a point of destructive interference.

The difference in the distances from the sources to point P or point Q is called the **path difference**. When a **whole number** of wavelengths can fit into this path difference, there is **constructive** interference at that point (producing a **maximum**), i.e. when the $\boxed{\text{path difference} = n\,\lambda}$ there is constructive interference (where n = 0 or 1 or 2 or 3 etc.).

When there is an odd number of half wavelengths in the path difference there is destructive interference at that point (producing a minimum), i.e. when the $\boxed{\text{path difference} = (n + \tfrac{1}{2})\,\lambda}$ there is destructive interference (where n = 0 or 1 or 2 or 3 etc).

Phase, Path Difference and Interference

Example

S₁ •————250 mm————• P Path difference to
P = 340 − 250 = 90 mm

S₂ •————340 mm

Whether constructive interference or destructive interference occurs at point P depends on both the wavelength and the path difference.

For a wavelength of 30 mm there would be three extra waves in the path from S_2 to P. This would make P a point of constructive interference. However, for a wavelength of 60 mm there would be 1.5 extra waves in the path from S_2 and P would be a point of destructive interference.

Gratings

A grating consists of a **large number** of **narrow slits** which are very close together and evenly spaced. When parallel light hits a grating, light spreads out (due to diffraction) from each of the slits and the slits act as a large number of **coherent sources**. The light diffracted from all the slits then overlaps and produces an interference pattern. This pattern is brighter and more clearly defined than the interference pattern made using only two slits.

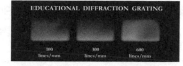

EDUCATIONAL DIFFRACTION GRATING

100 lines/mm   300 lines/mm   600 lines/mm

Monochromatic light consists of a **single wavelength** (or single frequency) – for example, red light from a laser. (White light from a lamp is not monochromatic – it is made up of all the many colours in the visible spectrum mixed together.) When monochromatic light passes through a grating, a simple interference pattern is produced. The points of constructive interference lie along lines centred on the middle of the grating.

Top Tip

Bigger wavelength, bigger spread.

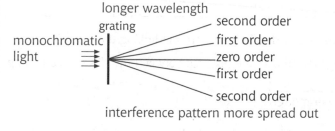

longer wavelength
grating
monochromatic light
second order
first order
zero order
first order
second order
interference pattern more spread out

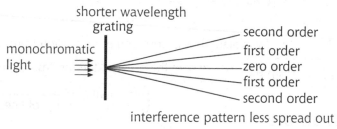

shorter wavelength
grating
monochromatic light
second order
first order
zero order
first order
second order
interference pattern less spread out

The spectral lines are all the same colour as each other (the same as the colour of the light source). The zero order spectrum is a **horizontal line of symmetry**. Between the coloured lines destructive interference takes place and there is **darkness**. Using a different grating with slits which are **closer** together causes the interference pattern to be **more spread out**.

Quick Test 27

1. Explain what is meant when waves are said to be 'in phase'.
2. Some waves combine in phase.
   (a) Describe what happens.
   (b) Name this effect.
3. Two wave sources, S_1 and S_2, produce identical, coherent waves. The wavelength of these waves is 30 mm. What happens at P?
4. Two wave sources, S_1 and S_2, produce identical, coherent waves. Point Q is the second minimum out from the central maximum. Calculate the wavelength of these waves.
5. Monochromatic light is passed through a grating to form an interference pattern on a screen. Make a list of the changes which would cause the interference pattern to be more spread out.

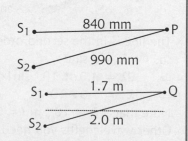

S₁ •————840 mm————• P
S₂ •   990 mm

S₁ •————1.7 m————• Q
S₂ •   2.0 m

Spectra using Gratings and Prisms

White Light Spectra

When **white light** passes through a grating a more complex interference pattern is observed.

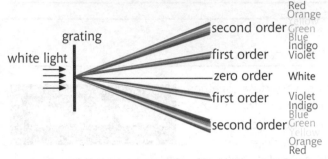

White light is made up of a whole range of different wavelengths of light mixed together. All the wavelengths meet in phase along the middle line and so the zero order spectrum is also white. Each different colour has a different wavelength and so the **different colours** meet in phase at **different positions** (because path difference must = nλ for constructive interference). As a result, the non-zero order spectra are spread out into the rainbow of colours we call the **visible spectrum**. The **red** ends of the spectra are always at a **greater angle** out from the central line than the violet ends.

The grating formula

When light of wavelength λ is observed at an angle θ from the zero order line, the formula linking the quantities is:

$$n\lambda = d \sin \theta$$

where n = the order of the spectrum being viewed d = the distance between the slits on the grating

$$\left(\text{Note that } d = \frac{1}{\text{no. of lines per metre}} \right)$$

Example
Light is incident on a grating which has 250 lines (= slits) per millimetre. A bright line is observed in the second order spectrum at an angle of 14.5°. What is the wavelength of the light?

250 lines per millimetre = 250 × 1000 lines per metre

so, d = 1/250,000 = 4.0 × 10⁻⁶ m

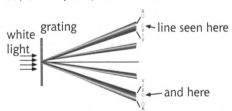

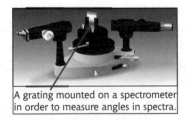

A grating mounted on a spectrometer in order to measure angles in spectra.

The line is in the second order spectrum, which means that 'n' = 2.

so, $n\lambda = d \sin\theta$

⇒ $2\lambda = 4.0 \times 10^{-6} \sin 14.5°$

⇒ $\lambda = 5.0 \times 10^{-7}$ m

Note that 5.0×10^{-7} m (= 500 nm) is the approximate wavelength of green light.

Other wavelengths you need to know are:

     red light – approximately 650 nm (6.5×10^{-7} m);

     blue light – approximately 480 nm (4.8×10^{-7} m).

Comparing spectra produced by gratings and triangular prisms

A triangular prism produces a **visible spectrum** from a ray of white light because different wavelengths of light **refract** different amounts.

The amount of refraction depends on the frequency of the incident light, i.e.

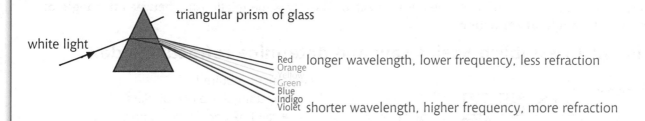

triangular prism of glass

white light

Red
Orange — longer wavelength, lower frequency, less refraction

Green
Blue
Indigo
Violet — shorter wavelength, higher frequency, more refraction

A grating produces spectra due to different wavelengths **interfering** constructively at slightly different positions (except for the zero order spectrum), i.e.

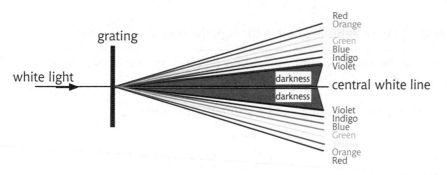

grating

white light

Red
Orange
Yellow
Green
Blue
Indigo
Violet

darkness
darkness — central white line

Violet
Indigo
Blue
Green
Yellow
Orange
Red

Top Tip

Remember the order of the colours using **ROY G BIV**.

- Only one spectrum is produced when using a triangular prism, whereas a grating produces an odd number of spectra. (e.g. 3, 5. [This is counting the zero order spectrum too.])
- Using a prism, the violet end of the spectrum is deviated most from the original direction, whereas with the grating, the red end of the spectrum is deviated most.
- The zero order spectrum produced by a grating is not spread out into a range of colours but always looks the same colour as the source appears to the naked eye.

1. White light is incident on a grating. An observer views the first order spectrum.
   (a) Which colour of light is seen at the greatest angle out from the central (zero order) maximum?
   (b) Is this the largest or smallest wavelength in the spectrum?
   (c) Is this the highest or lowest frequency in the spectrum?

2. Light is incident on a grating. The distance between the slits on the grating is 2.0×10^{-6} m. A bright line is observed in the second order spectrum at an angle of 28.7°.
   (a) What is the wavelength of the light?
   (b) What is the colour of the light?

3. A beam of white light is incident on a triangular prism. An observer views the spectrum.
   (a) Which wave property causes the light to spread into this spectrum?
   (b) Which colour in the spectrum has been deviated most from its original direction?

4. List the colours of the visible spectrum from high frequency to low frequency.

Refraction: Snell's Law

When does refraction occur?

Refraction occurs when light travels from one substance (medium 1) into another (medium 2). When refraction takes place, there is a **change in velocity** but **no change** in frequency. There is usually a change in direction – the exception is when light travels straight along the normal. Snell's Law is used to calculate the precise direction taken by a ray of light when it is refracted. The law is the relationship between the **angle of incidence** and the **angle of refraction**.

Experiment to establish Snell's Law and determine refractive index

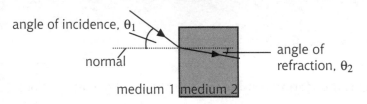

Equipment required:
- rectangular block of glass,
- ray box and power supply,
- protractor and ruler.

Experimental results produce this graph.

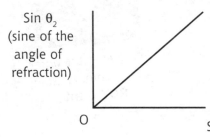

The graph (since it is a straight line through the origin) proves that;

> the sine of the angle of refraction is directly proportional to the sine of the angle of incidence.

This is called Snell's Law and can be written as the formula:

$$n = \frac{\sin \theta_1}{\sin \theta_2}$$

The '**n**' in the formula is called the **refractive** index of the glass block – it is a measure of how great the change is in velocity and direction when the ray of light enters the glass block. It has no units. When medium '1' is a **vacuum** (or air as an approximation) 'n' is called the **absolute** refractive index of medium '2'.

Many students find it useful to use Snell's Law in this form: $n_1 \sin\theta_1 = n_2 \sin\theta_2$

- Subscript '1' always refers to the medium the light is in and subscript '2' refers to the medium that the light is going to.
- The refractive index of a vacuum (and, approximately, for air) = 1.00.
- Angles are measured between the ray and the normal.

Top Tip

This relationship is not listed in the Physics data booklet, but it can be learned and used in the examination.

Example

A ray of light travelling in air is incident on the face of a glass block at an angle of incidence of 35°. The refractive index of the glass is 1.50. Calculate the angle of refraction inside the glass.

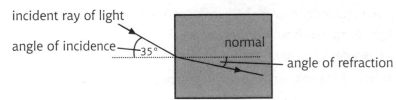

$$n_1 \sin\theta_1 = n_2 \sin\theta_2$$

data: n_1 = refractive index of air = 1.00 n_2 = refractive index of glass = 1.50

$\theta_1 = 35°$ and θ_2 = angle to be calculated

$\Rightarrow \quad 1.0 \times \sin 35° = 1.50 \times \sin\theta_2$

$\Rightarrow \quad \sin\theta_2 = 1.0 \times \sin 35°/1.50$

$\Rightarrow \quad \sin\theta_2 = 0.3824$

$\Rightarrow \quad \theta_2 = 22.5°$

Speed and wavelength on refraction

For a ray of light in medium 1 refracting into medium 2

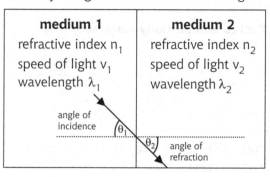

medium 1	**medium 2**
refractive index n_1	refractive index n_2
speed of light v_1	speed of light v_2
wavelength λ_1	wavelength λ_2

angle of incidence θ_1

θ_2 angle of refraction

Both the speed, v, and the wavelength, λ, of the light **change** when refraction occurs, but the **frequency** remains **constant**.
The speed of light in air/vacuum is 3.0×10^8 m s^{-1}.

The general relationship between the quantities is:

$$\frac{\sin\theta_1}{\sin\theta_2} = \frac{\lambda_1}{\lambda_2} = \frac{v_1}{v_2} = \frac{n_2}{n_1}$$

<u>frequency is constant</u>

Top Tip

These quantities may be matched in any pairings, for example the ratio for wavelengths equals the ratio for speeds.

Quick Test 29

1. Name the quantity which never changes on refraction.

2. A ray of light in air is incident on a glass block at an angle of incidence of 53°.

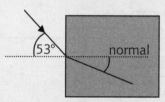

53° normal

The refractive index of the glass is 1.60 for this light. Calculate the angle of refraction.

3. A ray of light, which has a wavelength of 480 nm in air, is incident on a block of material as shown.

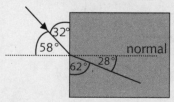

32° / 58° normal
62° / 28°

Use information from the diagram to calculate:
(a) the refractive index of the material for this light;
(b) the wavelength of the light inside the material;
(c) the frequency of the light in the material.

Total Internal Reflection and the Critical Angle

Internal reflection

When a ray of light is in a denser medium (for example, glass) and is incident on the interface with a less dense medium (e.g. air), it is possible that the ray will not refract out of the material. The ray is then **totally internally reflected**.

(a) When the angle of incidence, θ, is **less than** the critical angle, θ_c:

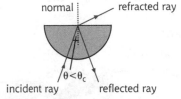

The incident ray is **partially reflected** and **partially refracted**. This is not total internal reflection.

(b) When the angle of incidence, θ, is **equal to** the critical angle, θ_c:

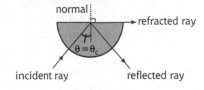

The ray is again **partially reflected**. The refracted ray emerges along the surface of the material, (i.e. the angle of refraction = 90°). This is not total internal reflection.

(c) When the angle of incidence, θ, is **greater than** the critical angle, θ_c:

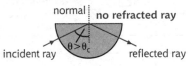

The ray is **100% reflected** inside the material (and obeys the law of reflection). This is total internal reflection.

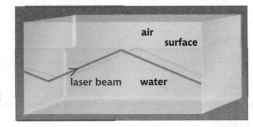

Snell's Law and total internal reflection

The **critical angle** is the value of the angle of incidence when the angle of refraction is **90°**, i.e.

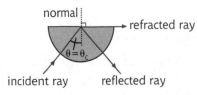

n_1 = refractive index of medium = n
n_2 = refractive index of air = 1.00
$\theta_1 = \theta_c$
$\theta_2 = 90°$

Substituting the values for this situation into Snell's Law gives:

$$n_1 \sin \theta_1 = n_2 \sin \theta_2$$
$$n \times \sin \theta_c = 1.00 \times \sin 90°$$
$$= 1.0$$

this gives $\boxed{n = \dfrac{1}{\sin \theta_c}}$ and $\boxed{\sin \theta_c = \dfrac{1}{n}}$

Top Tip

The relationship is only shown this way round in the Physics data booklet.

Example

The critical angle of a block of glass is found by experiment to be 42°.

Calculate the refractive index of the glass for this light.

$$n = \frac{1}{\sin \theta_c}$$

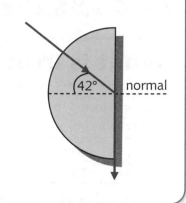

$\Rightarrow \qquad n = 1/\sin 42°$

$\Rightarrow \qquad n = 1/0.669$

$\Rightarrow \qquad n = 1.49 \qquad$ (Remember, refractive index has no units.)

Uses of total internal reflection

Total internal reflection causes light to **repeatedly reflect** off the inside surfaces of optical fibres. Optical fibres are used in many ways:

- to carry light from a lamp to illuminate instruments on the dashboard of a car;
- to carry telephone signals and other data at very high speed over long distances;
- to carry light in endoscopes to allow doctors to see inside patients.

Total internal reflection is used in 'cat's eyes' and reflective road signs. Total internal reflection is also used in **optical instruments** such as periscopes and binoculars where prisms are used to control the path taken by light.

Top Tip

Do some research to find more detailed information on uses of optical fibres.

Periscopes being used to see over a crowd.

periscope:

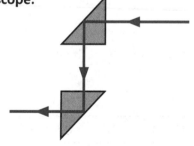

binoculars:

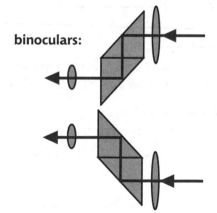

1. Write a sentence which describes the meaning of the term 'critical angle'.
2. What is the relationship between the angle of incidence and the critical angle when total internal reflection occurs?
3. A ray of light is in water (refractive index = 1.33). Calculate the critical angle at an interface of water with air.
4. A ray of light in diamond meets an interface with air as shown. Is the ray totally internally reflected? (The refractive index of diamond is 2.42.)
5. Reflected images have 'left' and 'right' swapped over (e.g. just as when you look at yourself in a mirror). Explain why the image seen through a periscope is the correct 'way round'.

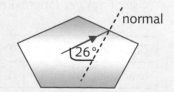

Irradiance

What is irradiance?

The irradiance, I, of light (or of any radiation) on a surface is defined as the power per unit area incident on that surface, i.e.

$$\text{irradiance} = \frac{\text{power}}{\text{area}} \quad \begin{array}{l} \text{— watts} \\ \text{— m}^2 \end{array} \qquad I = \frac{P}{A}$$

The units of irradiance are watts per square metre ($W\,m^{-2}$).

Example

A rectangular surface of dimensions 2.5 m by 4.5 m receives 171 joules of light energy every 4.0 seconds. Calculate the average irradiance of light on the surface.

Answer

   power = energy/time = 171/4.0 = 42.75 watts
   area = length × breadth = 2.5 × 4.5 = 11.25 m^2
   I = P/A = 42.75/11.25 = 3.8 $W\,m^{-2}$

Experiment

Experiment to find the relationship between irradiance and the **distance** from a **point source** of light.

Apparatus

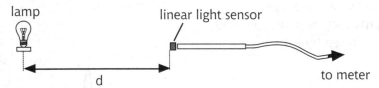

lamp

linear light sensor

to meter

> A small filament lamp is a good approximation for a point source of light.

Procedure

The linear light sensor is used to measure the irradiance of the light at various distances from the lamp (considered to be a point source of light).

Typical results

Distance (m)	0.20	0.40	0.60	0.80	1.00	1.50	2.00
Irradiance ($W\,m^{-2}$)	0.40	0.50	0.22	0.125	0.080	0.036	0.020

As distance, d, increases, irradiance, I, decreases.

> Irradiance is **inversely** proportional to the square of the distance from a point source.

Top Tip

Not all sources of light are point sources – for example, light from a LASER does not obey this relationship.

This relationship between I and d can be proved two ways.

(a) By calculation
   Multiplying together irradiance and the square of the distance gives a constant answer (approximately).

Distance (m)	0.20	0.40	0.60	0.80	1.00	1.50	2.00
Irradiance ($W\,m^{-2}$)	2.00	0.50	0.22	0.125	0.080	0.036	0.020
I d²	0.080	0.080	0.079	0.080	0.080	0.081	0.080

(b) A straight line graph through the origin is produced when irradiance is plotted against the inverse of the **square** of the distance, i.e.

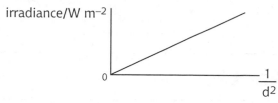

This means that doubling the distance decreases the irradiance to a quarter and tripling the distance decreases the irradiance to a ninth, etc.

Conclusion $\boxed{I = \dfrac{k}{d^2}}$

where 'k' is a constant.

Top Tip

This formula is not listed in the data booklet, but you can remember it and use it in the exam.

It is often convenient to use the following equation when carrying out calculations involving irradiance and distance from a point source.

$$\boxed{I_1 d_1^2 = I_2 d_2^2}$$

Where I_1 = the initial irradiance at a distance of d_1 from the point source,

I_2 = the final irradiance at a distance of d_2 from the point source.

Example

The irradiance of light at a distance of 0.50 m from a small filament lamp is measured to be 0.50 W m^{-2}.

Calculate the irradiance at a distance of 2.5 m from the lamp.

Answer

$I_1 d_1^2 \quad = I_2 d_2^2$

$0.5 \times (0.5)^2 = I_2 \times (2.5)^2$

$0.125 \quad = I_2 \times 6.25$

$I_2 \quad = 0.020$ W m^{-2}

Top Tip

You must remember to **square** the values of distance.

1. Light of power 6.0 W is incident on a surface of area 1.5 m^2.
   Calculate the irradiance of light on the surface.
2. A rectangular surface of dimensions 0.50 m by 1.60 m receives 280 joules of light energy every 7.0 seconds. Calculate the average irradiance of light on the surface.
3. An experiment is carried out to measure irradiance at different distances from a source of light. The results are as follows:

Distance (m)	0.10	0.30	0.50	0.70	0.90	1.10	1.30
Irradiance (W m^{-2})	5.0	0.56	0.20	0.10	0.062	0.041	0.030

   Can this source of light be considered to be a point source?
4. Sketch the graph which proves the relationship between irradiance and distance from a point source of light.
5. The irradiance of light at a distance of 0.75 m from a point source is 0.12 W m^{-2}.
   Calculate the irradiance at a distance of 1.5 m from the source.

The Photoelectric Effect 1

What is the photoelectric effect?

The photoelectric effect is the use of 'light' to **release electrons** from the **surface** of a metal. To demonstrate this effect:

- the metal used is normally **zinc**
- the metal must be **negatively** charged
- the radiation must have a higher frequency than visible light, e.g. **ultraviolet**.

Experiment

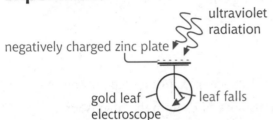

negatively charged zinc plate

ultraviolet radiation

gold leaf electroscope — leaf falls

> The instant the radiation is shone on to the surface of the metal, the gold leaf starts to fall. This shows that the negatively charged metal is discharging by losing excess electrons.

The theory that light is a wave cannot explain these results. (For example, the wave theory predicts that the metal cannot discharge immediately – a significant time should be required for enough energy to be supplied to release electrons.) The wave theory is not adequate so, a new theory is needed. This new theory, called the quantum theory, says that, in the photoelectric effect, light behaves as a stream of **individual energy bundles** called **photons.** The energy carried by a photon depends on the frequency of the radiation – the higher the frequency, the greater the energy.

> Photon energy, $E = h f$

where f = frequency of the radiation (in hertz); h = Planck's constant ($= 6.63 \times 10^{-34}$ J s)

When light is incident on the surface of the metal, **one photon** collides with **one electron** and gives up all its energy to that electron. There is a minimum quantity of energy an electron needs before it can escape from the attractive forces of the metal: this is called the **work function** of the metal. When the photon's energy is equal to or greater than the work function, the electron escapes from the metal and the plate discharges – the released electron is often called a photoelectron.

1.
radiation is shone on to plate

incident photon

metal plate

electron   metal atom

2.
a photon collides with an electron

3.
the electron escapes

released electron (photoelectron)

Example

The work function of a certain metal is 7.5×10^{-19} J.
The surface of the metal is now illuminated by radiation of wavelength 250 nm.
Are photoelectrons released from the surface of the metal?

Answer

Frequency of radiation, $f = v/\lambda = 3.0 \times 10^8/250 \times 10^{-9} = 1.2 \times 10^{15}$ Hz.
Photon energy $= hf = 6.63 \times 10^{-34} \times 1.2 \times 10^{15} = 7.96 \times 10^{-19}$ J
(This is greater than the work function).
The photon energy is greater than the work function. Therefore the electrons receive more energy than is needed to escape and so, yes, photoelectrons are released.

Top Tip

Remember, one photon gives all its energy to one electron.

The threshold frequency, f_o

When a source of light is used whose photons have less energy than the work function (e.g. using visible light), then electrons do not receive enough energy to escape from the metal and the **plate does not discharge**. There is a **minimum frequency** of incident light which causes photoemission – this is called the **threshold frequency**, f_o.

Photoemission does not occur when the frequency, f, of the radiation is below the threshold frequency, f_o. The threshold frequency depends on which metal is used because different metals have different values of work function, i.e. **f_o depends on the nature of the surface**.

Because the work function is equal to the minimum energy a photon needs to provide for an electron to be released, it can be calculated from:

$$\text{work function} = h\,f_o$$

When the photon energy is greater than the work function, electrons receive enough energy to escape from the metal. The plate then discharges and the released electrons possess **kinetic energy**. The maximum quantity of kinetic energy a photoelectron can possess is equal to the difference between the photon energy and the work function.

$$\text{maximum kinetic energy of a photoelectron} = \text{photon energy} - \text{work function}$$ or $E_k = hf - hf_o$

For hf less than hf_o
low energy photon

metal surface

electron   metal atom
photon energy is too small
to release an electron
NO PHOTOEMISSION OCCURS

For hf greater than hf_o
electrons released with kinetic energy
high energy photon

electron   metal atom
photon energy is greater than that
needed to release an electron
PHOTOEMISSION DOES OCCUR

Top Tip

Even with a high irradiance, there is no photoemission when f is less than f_o.

Example

The work function of a metal is 5.80×10^{-19} J.
Radiation of frequency 1.14×10^{15} Hz is incident on the metal.

(a) Calculate the threshold frequency of the radiation required for photoemission to occur.

(b) Calculate the maximum kinetic energy of the photoelectrons emitted from the surface of the metal with this radiation.

Answer

(a) $f_o = \text{work function}/h = 5.80 \times 10^{-19}/6.63 \times 10^{-34} = 8.75 \times 10^{14}$ Hz

(b) photon energy $= h\,f = 6.63 \times 10^{-34} \times 1.14 \times 10^{15} = 7.56 \times 10^{-19}$ J

maximum E_k = photon energy – work function = 7.56×10^{-19} J – 5.80×10^{-19} = 1.76×10^{-19} J

Quick Test 32

1. Monochromatic light of frequency 5.6×10^{14} Hz is incident on a surface. Calculate the energy of each photon.
2. Light of wavelength 540 nm is incident on a surface. Calculate the energy of each photon.
3. Light of frequency 7.9×10^{14} Hz is incident on the surface of a metal. The work function of the metal is 3.4×10^{-19} J. Calculate the maximum kinetic energy of the emitted photoelectrons.

The Photoelectric Effect 2

Increasing the irradiance

Increasing the irradiance means that **more photons per second** are incident on the surface. However, **each photon** still has the **same** quantity of **energy** as before (because $E = hf$, where h is a constant and the frequency, f, has not changed).

irradiance is low
few photons
per second

metal surface

irradiance is high
many photons
per second

metal surface

Two different situations need to be understood.
1. Increasing irradiance when the frequency of the radiation is **below** the threshold frequency.
2. Increasing irradiance when the frequency of the radiation is **above** the threshold frequency.

Below the threshold frequency

When the frequency of the radiation is below the threshold frequency, there is **no photoemission,** no matter how great the value of the irradiance. This is because although there are more photons per second, the energy of each photon is still the same (and since that **energy is below hf_o** – no photoelectrons are released).

For hf less than hf_o

low irradiance

metal surface

electron   metal atom
photon energy is too small
to release an electron
NO PHOTOEMISSION OCCURS

high irradiance

electron   metal atom
photon energy is still too small
to release an electron
NO PHOTOEMISSION OCCURS

Top Tip

Increasing the irradiance does not change the photon energy.

Above the threshold frequency

When the frequency of the radiation is above the threshold frequency, **more electrons** are released per second than before (i.e. the photoelectric current increases) but the **kinetic energy** of each electron has **not changed**.

For hf greater than hf_o

low irradiance

released electron

a FEW electrons are released per second
PHOTOELECTRIC CURRENT IS LOW

high irradiance

MANY electrons are released per second
PHOTOELECTRIC CURRENT IS HIGH

The photoelectric current is directly proportional to the irradiance of the light (above the threshold frequency).

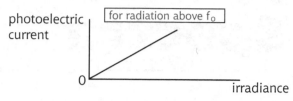

photoelectric current

for radiation above f_o

0

irradiance

Irradiance, I, and the number of photons per second per unit area, N

Consider a sheet of metal of area 1.0 m² which is receiving photons at a rate of N photons per second.

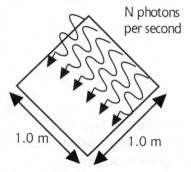

N photons per second

1.0 m   1.0 m

Top Tip

N is not just a number. It is the number per second per square metre.

Each photon has energy = hf.
The total energy incident on the surface per second must be Nhf.
But, of course, energy per second = power.
Irradiance = Power/Area = Nhf/1 = Nhf
so,

irradiance, $I = Nhf$

The wave-particle duality

In some experiments, such as gratings producing interference patterns, light behaves as **waves**. In other experiments, such as the photoelectric effect, the results can only be explained by light travelling as a stream of particles (**photons**). There is **no single theory** to totally explain the behaviour of light. At present, we have to retain **both theories** and use each of them in different circumstances. This is called the **wave-particle duality**.

Quick Test 33

1 Describe what change occurs when the irradiance of light on a surface is increased.
2. Explain why it is possible for irradiance to be increased when each photon still has the same energy as before.
3. Explain why it is possible for no photoemission to occur even when the irradiance of light on the surface of a metal is very high.
4. Why does an increase in irradiance not cause emitted electrons to have increased kinetic energy?

Atomic Energy Levels and Emission Spectra

The Atomic Model

Our model of the atom is of a **tiny compact nucleus** surrounded by **orbiting electrons.** These electrons can only occupy certain orbits (or **discrete energy states**).

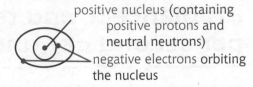

positive nucleus (containing positive protons and neutral neutrons)

negative electrons orbiting the nucleus

When a satellite is orbiting the earth, more energy is needed to lift the satellite into a higher orbit. Similarly, **higher electron orbits** are equivalent to **higher energy states**. Energy must be supplied to raise an electron from a lower state into a higher energy state (= higher orbit). Energy is given out again when an electron falls from a higher state to a lower one.

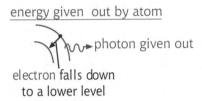

energy put in to atom

photon put in

electron moves up to a higher level

energy given out by atom

photon given out

electron falls down to a lower level

Rather than showing curved lines representing orbits, diagrams are usually drawn with horizontal lines representing energy levels. In a **free atom** of the simplest substance, **hydrogen**, there is a **single electron** which has a **limited number** of allowed orbits. These are represented by the following **energy level diagram**.

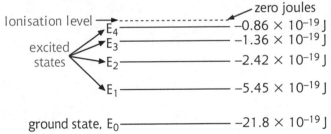

zero joules

Ionisation level

E_4 —————— -0.86×10^{-19} J

excited states

E_3 —————— -1.36×10^{-19} J

E_2 —————— -2.42×10^{-19} J

E_1 —————— -5.45×10^{-19} J

ground state, E_0 —————— -21.8×10^{-19} J

The **ground state** corresponds to the lowest orbit, i.e. the one closest to the nucleus. This is where the electron is normally found in a hydrogen atom. When an atom of hydrogen is supplied with extra energy (for example, by passing an electric current through the gas) the electron may be raised to any one of the higher energy levels (called **excited states**). The electron is still held by the attractive forces of the nucleus, i.e. it is bound to the atom. Once an electron has been given sufficient energy to reach the ionisation level it is freed from the atom. The **ionisation level** is used for the **zero** level of **energy** values.

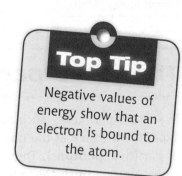

Top Tip

Negative values of energy show that an electron is bound to the atom.

Example 1

The electron in an atom of hydrogen is in the ground state.
How much energy must be given to the electron to free it from the nucleus?

Answer

The electron must be given enough energy to raise it from the ground state to the ionisation level. The energy required = the size of the **gap** between E_0 and the ionisation level = 21.8×10^{-19} J.

Example 2

The electron in an atom of hydrogen is in the ground state.
How much energy must be given to the electron to raise it to level E_1?

Answer

The energy required = the size of the gap between E_0 and E_1 = $(21.8 \times 10^{-19} - 5.45 \times 10^{-19})$ J
$$= 16.35 \times 10^{-19} \text{ J}$$

Emission spectra

The emission spectrum for a substance is the range of colours seen when the light it emits is passed through a grating (or prism). An electron does not remain in an excited state for long. It soon **falls** to a lower level and **emits** a **photon**. This photon possesses an amount of energy equal to the **gap** between the levels, for example,

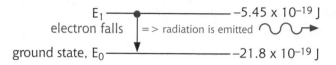

E_1 ———•——————— -5.45×10^{-19} J

electron falls $=>$ radiation is emitted

ground state, E_0 ———————— -21.8×10^{-19} J

Top Tip

Don't let the negative signs put you off. Just find the size of the gap.

Here, the energy of the emitted photon is $(21.8 - 5.45) \times 10^{-19}$ J $= 16.35 \times 10^{-19}$ J. The emitted energy is often in the form of visible light. For a multi energy level system there are **several** possible **transitions** through which electrons can fall.

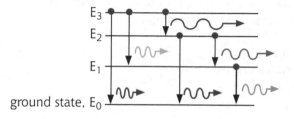

E_3

E_2

E_1

ground state, E_0

The transition from E_3 to E_0 produces the largest energy output; this corresponds to a photon of **high frequency** (= short wavelength) – this is at the **violet** end of the spectrum. The transition from E_3 to E_2 produces the **smallest energy** output; this corresponds to a photon of **low frequency** (= long wavelength) – this is at the **red** end of the spectrum.

The emission spectrum for the above transitions is:

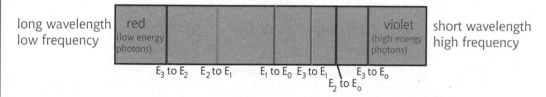

long wavelength / low frequency — red (low energy photons) ... violet (high energy photons) — short wavelength / high frequency

E_3 to E_2 E_2 to E_1 E_1 to E_0 E_3 to E_1 E_3 to E_0
E_2 to E_0

> Each line in the emission spectrum is due to an electron making a different transition from an excited energy level to a lower level.

Every different line in the spectrum corresponds to the emission of photons of a different frequency. Since the atom of any particular element has a unique pattern of energy levels, each element produces a unique set of frequencies when light is radiated from it. This explains why each element has a **different line spectrum**.

Note the following statements carefully.

1. A **bright line** in an emission spectrum is because **many electrons** make that transition each second.

2. A transition through a **large gap** between levels produces a photon of **high energy** (e.g. violet light), but when only a few electrons make that transition per second the **violet** line is dim.

Quick Test

This Quick Test is on page 73 – try questions 1 to 4.

Absorption Spectra

When an atom **absorbs** energy, an electron is excited and moves **up** from a lower energy level to a higher energy level.

higher energy level, W_2 ─────────────

radiation is absorbed

=> electron moves up to higher level

lower energy level, W_1 ─────────────

The electron only makes this upward transition when the energy of the incident photon is **equal to the energy gap.**

This means that $\boxed{W_2 - W_1 = hf}$ where f is the frequency of the radiation being absorbed.

When light containing a range of frequencies is shone through these atoms, only photons of the **correct frequency** will be **absorbed**. Photons having different frequencies will not be absorbed.

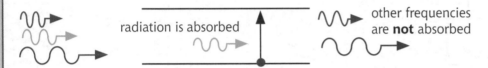

radiation is absorbed

other frequencies are **not** absorbed

As a result the spectrum of the **remaining light** will show a dark line corresponding to the missing frequency (or frequencies). This is called an absorption spectrum.

Observing the absorption spectrum of sodium

Here is one experimental arrangement for observing the absorption spectrum of sodium.

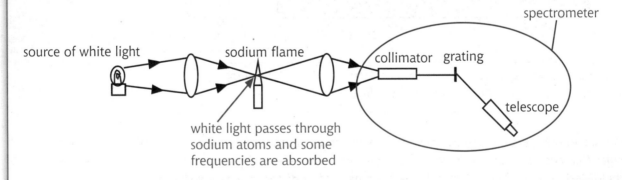

spectrometer

source of white light   sodium flame   collimator   grating

telescope

white light passes through sodium atoms and some frequencies are absorbed

Results

Absorption spectrum of sodium

black lines due to absorption of energy

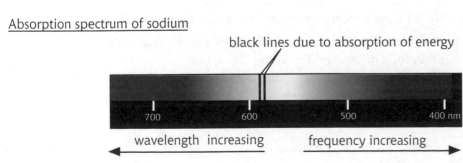

700   600   500   400 nm

← wavelength increasing   frequency increasing →

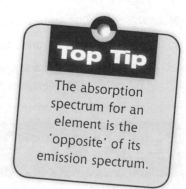

Top Tip

The absorption spectrum for an element is the 'opposite' of its emission spectrum.

The energy absorbed by the sodium atoms is **immediately re-radiated** by them as electrons fall back down to lower levels. This radiation is, however, emitted in **all directions** – so the amount of energy re-radiated in the original direction is very small, causing these frequencies to be 'missing' in the spectrum.

Absorption lines in the spectrum of sunlight

You probably expect that light from the Sun would produce a complete visible spectrum from red through to violet. However, when sunlight is carefully observed with a spectrometer, a number of fine dark lines are seen. These dark lines are due to some of the light emitted from the Sun being absorbed by cooler atoms in the outer layers of the Sun's atmosphere.

Quick Test 34

1. Describe what is happening in an atom when:
   (a) it absorbs energy;
   (b) it emits energy.

2. What name is given to:
   (a) the energy level corresponding to the lowest electron orbit;
   (b) energy levels above the lowest level;
   (c) the level where the electron has received enough energy to escape from the atom?

3. Look at the following energy level diagram.

   E_4 ———————————— -0.86×10^{-19} J
   E_3 ———————————— -1.40×10^{-19} J

   E_2 ———————————— -2.45×10^{-19} J

   E_1 ———————————— -5.55×10^{-19} J

   Calculate:
   (a) the energy required to raise an electron from level E_1 to level E_4;
   (b) the energy of the photon emitted when an electron falls from level E_4 to E_2;
   (c) the highest frequency of radiation emitted due to a transition between these levels;
   (d) the longest wavelength of radiation emitted due to a transition between these levels.

4. Explain why a line in an emission spectrum is brighter than the others.

5. Describe what is happening in the atoms of a substance when its absorption spectrum is formed.

6. The following diagram shows the absorption spectrum for a particular substance.

   Sketch the emission spectrum for the same substance.

7. Explain why there are faint dark lines in the spectrum of light from the Sun.

The Laser

To understand the workings of a laser, you must first know about the difference between **spontaneous** emission of radiation and **stimulated** emission of radiation.

Spontaneous emission

When an atom has already absorbed some energy, one of its electrons is in a higher energy level than normal. That electron will not remain in that higher level for long. Typically within a few milliseconds the electron falls back down from the higher level to a lower level causing the emission of a photon. The energy possessed by the photon is equal to the **energy difference** between the two energy levels.

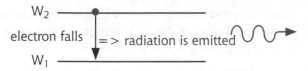

This is called **spontaneous emission** because there has been **no external influence** on the atom to cause or encourage the fall of the electron. It is impossible to predict exactly when a particular atom is going to radiate a photon by the fall of an electron because it is a **random process**. Spontaneous emission is therefore similar to the radioactive decay of a nucleus where it is impossible to predict when a particular nucleus will breakdown.

Top Tip

$W_2 - W_1 = hf$

Stimulated emission

An electron may be **caused** to fall from a higher energy level to a lower level by sending in an initial (or stimulating) photon which has exactly the same amount of energy as the gap between the two levels. The emitted photon is found to be both in phase and in the same direction as the initial photon, i.e. the stimulating and emitted photons are identical to each other.

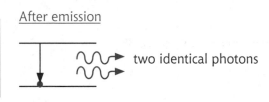

This is called **stimulated** emission because there has been an **external influence** on the atom to cause the electron to fall between the levels.

It is because of stimulated emission that it was possible to invent the laser in the 1950s.

In the examination you should be able to state that:
'When radiation of energy hf is incident on an excited atom, the atom may be stimulated to emit its excess energy hf.'

The laser

The name of this device comes from
Light **A**mplification by **S**timulated **E**mission of **R**adiation.

In a laser, electrons are continually 'pumped' up into higher energy levels by an external source of energy. These electrons are then stimulated to fall back down and emit photons. The photons then go on to stimulate the fall of **even more** electrons. Some of the photons are lost by absorption in the medium, but, overall the light beam gains **more energy** by stimulated emission than it loses by absorption – this is **amplification**.

The lasing medium (for example, a mixture of helium and neon gases) is contained between two **parallel mirrors**. These mirrors reflect the photons backwards and forwards through the medium, **stimulating further emissions**. One mirror is a 100% reflector. The other mirror is a **partial reflector**. It only reflects about **98%** of the light, the other **2%** of the light emerges as the **narrow laser beam**.

The laser diagrammatically

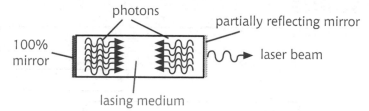

Laser light is:

    (a) **monochromatic**, because all the photons have the same frequency (and wavelength);
    (b) **coherent**, because all the photons are in phase;
and   (c) produces a **high irradiance**, because the power is concentrated on to a very small area.

LASER RADIATION
Laser warning sign

Example

A laser can have a power as low as 0.1 mW. This value of power is very small compared to a 60 W lamp, but it can be concentrated on to a spot as small as 0.5 mm in diameter.

Calculating the irradiance this produces on the illuminated surface:

    area = area of a circle = $\pi r^2 = 3.142 \times (0.25 \times 10^{-3})^2 = 1.96 \times 10^{-7}$ m^2
    irradiance = P/A = $0.1 \times 10^{-3}/1.96 \times 10^{-7} = 510$ W m^{-2}

This compares with an irradiance of approximately 5.0 W m^{-2} at a distance of 1.0 m from a 60 W lamp (and this is assuming that all the 60 W of power is in the form of light energy). It is therefore easy to understand why even a low power laser can cause eye damage.

An important point

It is important to remember that a laser is **not** a point source of light – the photons are all travelling in the same direction. As a result the **beam spreads out very little** and so the irradiance of the beam decreases very little with distance.

American astronauts left a mirror on the moon's surface in order to reflect a laser beam sent from Earth. The mirror is used in experiments to measure the distance to the moon very accurately. This experiment can only work because a laser beam spreads out very little.

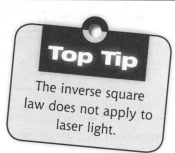

Top Tip

The inverse square law does not apply to laser light.

Quick Test 35

1. Describe any similarities and differences between the spontaneous emission of light and the simulated emission of light.
2. In what way is the spontaneous emission of light similar to the radioactive decay of a nucleus?
3. Which type of emission occurs inside a laser to make it work?
4. State ways in which the stimulating photon and the emitted photon are the same as each other.
5. Explain why even a low power laser can cause eye damage.

Conductors, Insulators and Semiconductors

Electrical conduction in solids

Conductors

Conductors are materials which have a significant number of **free electrons** in their structure. When an electric field is applied to such a substance the electrons experience a **force** towards the positive terminal and so **move** in that direction, i.e. an **electrical current** is produced.

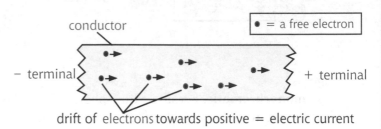

drift of electrons towards positive = electric current

Examples of good conductors are copper, silver and gold, i.e. metals, and the **graphite** form of carbon.

Insulators

Insulators are materials which do **not** have a significant number of free electrons in their structure (their electrons are needed for chemical bonding). When an electric field is applied to such a substance the electrons experience a force towards the positive terminal but are **unable to move** and so there is **no** electrical current.

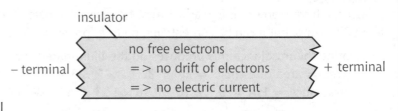

Examples of good insulators are wood, paper, plastic, rubber, glass etc. and the **diamond** form of carbon.

Semiconductors

There is another group of materials which are **neither** good conductors nor good insulators. These materials have a **small number** of charge carriers available for conduction and so are called **semiconductors**. The charge carriers are either **negative electrons** or **positive holes** (to be explained!).

Silicon   Germanium

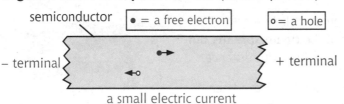

a small electric current

Top Tip

Be very careful to spell correctly!

Examples of semiconductors are silicon (not silicone!) and germanium (not geranium!).

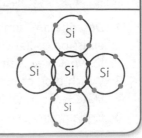

Geranium

Understanding more about charge carriers in semiconductors

Silicon (and germanium) atoms each have **four** electrons in their outer shells. Each silicon atom **shares** these electrons with its four nearest neighbours to form a diamond shaped structure and gives each atom a share in **eight electrons**.

The bonds holding the electrons in these positions are not very strong and only **a little energy** is sufficient to break some of them and set a **few electrons** free. Room temperature is sufficient to provide this energy. These free electrons are then available for electrical **conduction**.

Whenever an electron breaks free from the bonding structure a gap or **hole** is left behind.

This hole then behaves like a **positive charge carrier** and is available for conduction. A hole moves towards the negative terminal when an electron jumps towards the positive terminal.

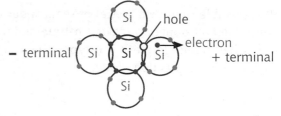

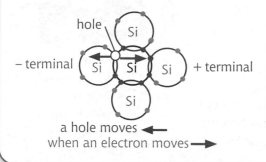

a hole moves ← when an electron moves →

This type of conduction happens in a pure semiconductor (and is called intrinsic conduction). As the temperature is increased, more bonds are broken and this intrinsic conduction increases. This is why the resistance of many thermistors decreases as their temperature increases.

Doping semiconductors

It is possible to increase the number of charge carriers available for conduction in a semiconductor by **adding** a small number of different (or **impurity**) atoms. This is called **doping** the semiconductor and it **decreases** the resistance of the material. For example, a few atoms of **arsenic** (chemical symbol 'As') can be introduced into a block of silicon. Each atom of arsenic has **five** electrons in its outer shell, but only four of these are needed for the bonding structure.

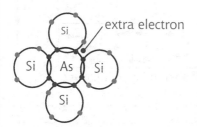

There is one extra electron per arsenic atom which is not needed for bonding. These extra electrons are **free for conduction**. In this type of semiconductor, the majority of the charge carriers are **negative** electrons and so it is called an **n-type** semiconductor. However, the block of material is **not** negatively charged; it is electrically neutral.

Top Tip

Impurity atoms with five electrons in the outer shell produce n-type.

Another example of doping is when a few atoms of **indium** (chemical symbol 'In') are introduced into a block of silicon. Each atom of indium has **three** electrons in its outer shell. There is now **one fewer** electron per indium atom than is needed for bonding and so **holes** are available for conduction. In this type of semiconductor the majority of the charge carriers are **positive** holes and so it is called a **p-type** semiconductor. However, the block of semiconductor is **not** positively charged; it is electrically neutral.

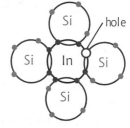

Quick Test 36

1. Name   (a) a good conductor
            (b) a good insulator
            (c) a semiconductor.
2. A block of silicon is doped to make a p-type semiconductor. How many electrons are in the outer shell of the impurity atoms?

The p-n Junction Diode

It is possible to grow a crystal of semiconductor with one half doped to make it **p-type** and the other half doped to make it **n-type**. The resulting crystal is a p-n junction diode.

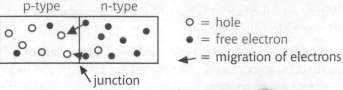

○ = hole
● = free electron
⟵ = migration of electrons

junction

A few electrons from the n-type material **migrate** across the junction to the p-type material and 'fill' the holes. As a result, there is a narrow band either side of the junction which has a **shortage** of charge carriers. This is called the **depletion layer**.

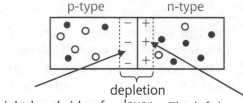

depletion layer

The right hand side of the p-type has become slightly negative.

The left hand side of the n-type has become slightly positive.

Top Tip

Note that there are some free electrons in the p-type and some holes in the n-type because of thermal energy.

This migration of charges has created a small **potential difference** across the junction which, because of electrostatic forces, **opposes** any further migration of charges. This depletion layer is very **thin**.

The forward biased p-n junction diode

When the p-n junction diode is connected to a d.c. power supply with the **p-type** material connected to the **positive** terminal, the forces on the charges cause the depletion layer to **collapse**. Charge flows **continuously** with electrons in the n-type drifting towards the junction and holes in the p-type also drifting towards the junction. The electrons and holes recombine at the junction allowing a steady current in the whole circuit. The diode is said to be **forward biased**.

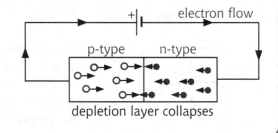

electron flow

p-type   n-type

depletion layer collapses

The reverse biased p-n junction diode

When the **p-type** material is connected to the **negative** terminal of a cell, the forces on the charges cause the depletion layer to **widen**. The semiconductor acts as an **insulator** and there is **no current** in the circuit. A p-n junction diode therefore allows charge to flow in only **one direction**. This is why it can be used to **rectify** a.c.

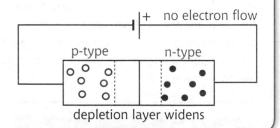

no electron flow

p-type   n-type

depletion layer widens

The p-n junction diode symbol

This is the symbol for the p-n junction diode.

You need to remember that electrons are allowed to flow this way relative to the symbol.

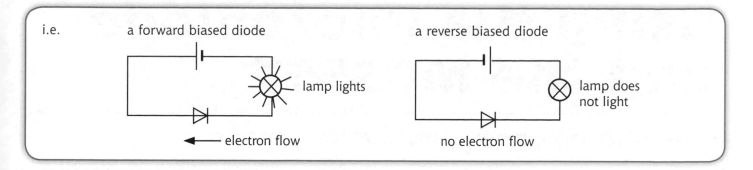

i.e.   a forward biased diode   a reverse biased diode

lamp lights   lamp does not light

← electron flow   no electron flow

The light emitting diode

The light emitting diode (or **LED**) is a semiconductor p-n junction which also **emits** light when it is conducting (i.e. when it is **forward biased**). The emission of light happens because a **photon** of light is emitted whenever an **electron** and a **hole recombine** at the junction.

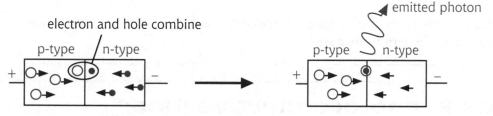

electron and hole combine

emitted photon

p-type   n-type   p-type   n-type

> **Top Tip**
>
> Don't confuse the combination of an electron and a hole in an LED with the separation of an electron and a hole in a photodiode.

The symbol for an LED is like that for a p-n junction diode, with the addition of two arrows pointing **away** to indicate the emission of light.

The photodiode

A photodiode is a p-n junction in a transparent package allowing it to **respond to light** incident on it. The symbol for a photodiode is like that for a p-n junction diode, with the addition of two arrows pointing **towards** it to indicate the incident light.

When the **photons** hit the junction they **separate** electrons from holes. This produces free charge carriers which are available for conduction.

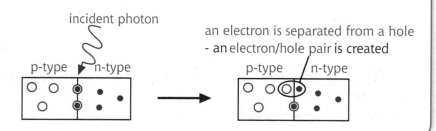

incident photon

an electron is separated from a hole - an electron/hole pair is created

p-type   n-type   p-type   n-type

Quick Test 37

1. Describe what is meant by the term 'depletion layer' in a p-n junction.
2. A p-n junction is connected to a d.c. supply so that it is forward biased. Describe how to do this.
3. Describe what happens in an LED to produce light.

Using the Photodiode and the MOSFET

The photodiode in photovoltaic mode

In the photovoltaic mode, the photodiode may be used to supply power to a load.

No other power supply is used – the photodiode converts light energy directly into electrical energy. The photodiode produces an e.m.f. which can be used to operate, for example, a motor.

When the **irradiance** of the light incident on the photodiode is **increased**, the number of electron-hole pairs produced each second increases and so the **e.m.f. increases**.

The photodiode in photoconductive mode

In the photoconductive mode, a **photodiode** is connected in series with a resistor and in **reverse bias** with a power supply.

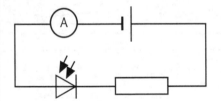

When there is **no light** incident on the photodiode, it is a very good **insulator** and there is **no current**. When light is incident on the photodiode, photons **generate electron-hole** pairs at the junction, the **resistance** of the photodiode **decreases** and there **is** a current in the circuit (sometimes referred to as the leakage current).

The number of electron-hole pairs is directly proportional to the number of photons hitting the junction. As a result of this, the leakage current is directly proportional to the irradiance of the radiation. This means that, in the photoconductive mode, the photodiode can be used as a linear light sensor. The photodiode responds extremely quickly to changes in light level.

MOSFET

The <u>M</u>etal <u>O</u>xide <u>S</u>emiconductor <u>F</u>ield <u>E</u>ffect <u>T</u>ransistor, (MOSFET).
Circuit symbol
The three terminals of the MOSFET are called:

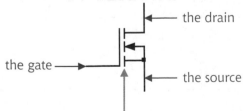

the drain

the gate

the source

Top Tip

You need to learn the symbol, the names of the terminals and be able to describe the structure of a MOSFET.

The vertical broken line in the symbol indicates that there is normally no conducting path between the source and the drain.

How MOSFETs are made

A piece of **p-type semi-conductor** is the starting point. This is called the **substrate**. Using a process called diffusion, two n-type regions are implanted at either end. These are called the **n region implants**. A thin layer of insulating oxide is now deposited on top of the substrate and implants. This is the **oxide layer.** In order to make electrical contact with the n region implants, some of the oxide layer is etched away. **Metal contacts** are made to the n region implants and to the insulating oxide layer between them. The oxide layer under the metal contact is very thin. A connection is made between the substrate and one of the n region implants.

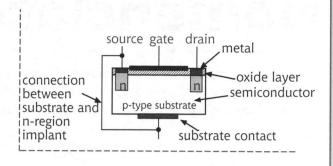

Switching on a MOSFET

A potential difference is applied between the gate and the source to make the gate **positive**.

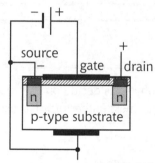

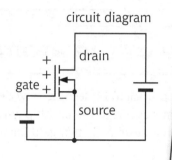

circuit diagram

Top Tip

Don't confuse the MOSFET with other types of transistor which switch on when the input voltage is 0.7 V.

The positive potential on the gate causes a **conducting layer** of electrons to form below the gate just beneath the insulator. This channel is called an **n-channel** (since it is formed by **n**egative electrons). Making the **drain positive** with respect to the source now causes electrons to **flow** from the source to the drain along the n-channel.

As long as the gate is approximately **2.0 V more positive** than the source, electrons can flow along the n-channel, i.e. the **MOSFET** is **switched on**. Whenever the potential of the gate falls **below +2.0 V**, the n-channel disappears, electrons cannot flow and the **MOSFET** is **switched off**. The n-channel MOSFET can also be used as an **amplifier**.

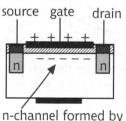

n-channel formed by electrons attracted to the positive gate

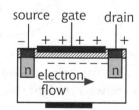

electrons flow from the source to the drain as long as the channel is formed.

Quick Test 38

1. Name the two modes in which a photodiode can be used.
2. A photodiode is in a circuit in which it is connected to a cell. In which mode is it being used?
3. In which mode can a photodiode be used as a linear light sensor?
4. Sketch a circuit diagram showing a photodiode in photovoltaic mode to operate a motor.
5. Explain why the conducting channel in a MOSFET is called an n-channel.
6. Sketch a diagram of an n-channel MOSFET connected so that it can be switched on (i.e. show the polarities of the power connections).

Atomic Structure and Nomenclature

Our model of the atom is that it consists of a **tiny compact nucleus** (in which positive protons and neutral neutrons are bound closely together) and around which negative **electrons** are **orbiting**. Between the nucleus and the closest electron orbit there is only **empty space**.

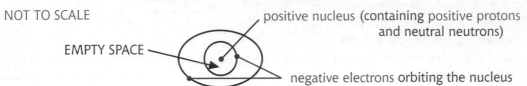

NOT TO SCALE

EMPTY SPACE

positive nucleus (containing positive protons and neutral neutrons)

negative electrons orbiting the nucleus

It cannot be stressed strongly enough that **this diagram is not to scale.** If the atom could be enlarged until the nucleus was about the size of a golf ball, there would be a distance of approximately **one kilometre** from the nucleus to the nearest orbiting electron!

The structure of the atom

The structure of the atom is too small to be seen; so how did physicists decide on this model? A major contribution to our present model of the atom was due to the British physicist, Ernest Rutherford. His experiment consisted of a beam of alpha particles fired towards an extremely thin gold foil.

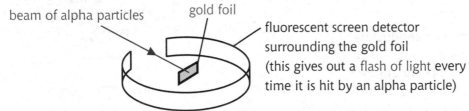

beam of alpha particles

gold foil

fluorescent screen detector surrounding the gold foil (this gives out a flash of light every time it is hit by an alpha particle)

Previous theories about the structure of atoms predicted that a reasonably high number of the alpha particles would be deviated a small amount from the straight through path but Rutherford's experimenters found that most of the alpha particles **passed straight through** the gold foil as if they had **not encountered** any matter at all.

Top Tip

This was the most surprising result of the experiment!

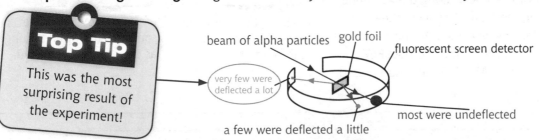

beam of alpha particles gold foil

fluorescent screen detector

very few were deflected a lot

most were undeflected

a few were deflected a little

These results led Rutherford to the conclusion that **most of the atom is empty space**.

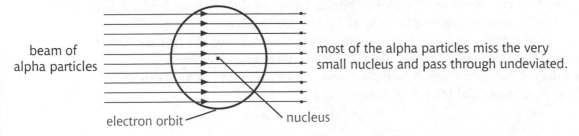

beam of alpha particles

most of the alpha particles miss the very small nucleus and pass through undeviated.

electron orbit

nucleus

The most surprising experimental result was that a very small number of the alpha particles were deflected back the way they had come. This was as astonishing as seeing an express train colliding with, and bouncing off a sheet of paper!

Rutherford concluded that the alpha particles had been repelled from something of much greater mass than themselves; this meant that most of the **mass** of the atom must be **concentrated** in a **very small volume**, which we now call the **nucleus**.

large deflection of an alpha particle after a collision with nucleus

undeviated alpha particles

nucleus

The small deflection of some of the alpha particles is due to 'near misses'.

path of alpha particle

nucleus

small deflection after passing very close to the nucleus

Nomenclature

The term **nucleon** means a **particle in the nucleus**. This means that both **protons** and **neutrons** are nucleons. Electrons are not nucleons because they are not in the nucleus; they only orbit around it. The **mass number, Z**, of an atom is the **total number** of its **nucleons**. This means that the mass number is the sum of the number of protons and the number of neutrons. The **atomic number, A**, of an atom is its the **number of protons**. In a normal, electrically neutral atom this is also equal to the number of electrons that orbit its nucleus.

The chemical symbol for the atom is written with the

mass number, Z, above and to the left
atom number, A, below and to the left.

$$^{Z}_{A}\text{Symbol}$$

Examples

$^{4}_{2}\text{He}$ ← An atom of helium which has two protons and two neutrons in its nucleus (and two orbiting electrons in a normal neutral atom).

Top Tip

The atomic number identifies the element.

$^{235}_{92}\text{U}$ ← An atom of uranium which has 92 protons and 143 neutrons in its nucleus (and 92 orbiting electrons in a normal neutral atom).

- The chemical **symbol** and the **atomic number** give the **same information**, i.e. it is not possible to have the symbol 'U' for uranium with an atomic number other than 92.
- You are not expected to memorise the symbols for the elements or to know their atomic numbers. All this information is provided in the **Periodic Table** inside the Physics data booklet.

Isotopes

Different atoms of the **same element** can have **different numbers of neutrons**, e.g. some atoms of the element uranium have 143 neutrons (and so a mass number of 235) and some other atoms have 146 neutrons (and so a mass number of 238), i.e.

$^{235}_{92}\text{U}$ and $^{238}_{92}\text{U}$

These are called **isotopes** of uranium.

Quick Test 39

1. Which of Rutherford's results suggested that the atom is mainly made up of empty space?
2. The symbol for an atom is $^{207}_{82}\text{Pb}$. Give details of the particles which make up this atom.

Radioactive Decay

Different types of radioactive decay

The different types of radioactive decay are:

    alpha (α) particles   beta (β) particles   and gamma (γ) rays

One of the ways of telling them apart is by how **penetrative** they are when passed through materials.

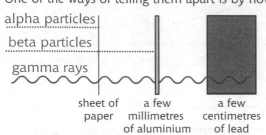

alpha particles

beta particles

gamma rays

    sheet of   a few   a few
    paper   millimetres   centimetres
    of aluminium   of lead

Alpha particles are stopped by a thin sheet of paper (or a few centimetres of air).

Beta particles are stopped by a few millimetres of aluminium.

Gamma rays are stopped by a few centimetres of lead (or a few metres of concrete).

All radioactive decay is due to **changes in the nucleus** of an atom resulting in the emission of particles or bursts of electromagnetic energy.

Radiation warning sign

Alpha radiation

An alpha (α) particle is made up of two protons and two neutrons joined together. When a nucleus emits an alpha particle it **loses two protons and two neutrons.** When the original nucleus (the **parent** nucleus) emits an alpha particle, the **atomic number decreases by two** and the **mass number decreases by four.** The remaining nucleus (called the **daughter** nucleus) is a **different element** because it has a different atomic number, e.g.

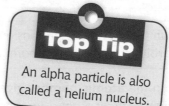

an α particle   key: ○ = a proton ● = a neutron

nucleons which will split off as the alpha particle

    parent nucleus   daughter nucleus   emitted alpha particle
    10 protons   8 protons
    8 neutrons   6 neutrons   key: ○ = a proton
    => mass no. = 18   => mass no. = 14   ● = a neutron
    atomic no. = 10   atomic no. = 8

Top Tip

An alpha particle is also called a helium nucleus.

Example

Uranium-238 emits an alpha particle. Give a statement for this decay.

Answer

$$^{238}_{92}U \rightarrow\ ^{234}_{90}Th + ^{4}_{2}\alpha$$

This is a pictorial illustration – in reality, this type of decay usually only occurs with heavier, unstable nuclei.

Beta radiation

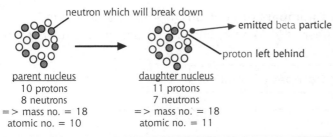

neutron which will break down

    → emitted beta particle

    proton left behind

    parent nucleus   daughter nucleus
    10 protons   11 protons
    8 neutrons   7 neutrons
    => mass no. = 18   => mass no. = 18
    atomic no. = 10   atomic no. = 11

A beta (β) particle is a **very fast moving electron** ejected from the nucleus. There are no electrons normally in the nucleus – a **neutron breaks down** into a proton and an electron. The proton remains behind in the nucleus and the electron is ejected as the beta particle,

This is a pictorial illustration – in reality, this type of decay only occurs with heavier, unstable nuclei.

Example

Thorium-234 emits a beta particle. Give a statement for this decay.

Answer $^{234}_{90}Th \rightarrow\ ^{234}_{91}Pa + ^{0}_{-1}\beta$

When beta decay occurs, the mass number of the daughter nucleus does not change and the atomic number **increases** by one. The symbol for a beta particle can be $^{0}_{-1}\beta$ or $^{0}_{-1}e$.

Gamma radiation

Gamma (γ) radiation is the emission of **short bursts** of **very high frequency** electromagnetic radiation from the nucleus due to rearrangement of the nucleons. As a model of this, you could compare a pyramid of marbles collapsing and giving out sound energy (from the original potential energy).

pyramid of marbles collapsing

sound energy emitted

The nucleus, in a similar way, can have its protons and neutrons in a high energy arrangement 'falling' into a lower energy arrangement and emitting energy as gamma radiation. There is therefore no change in the number (or type) of particles in the nucleus when gamma radiation is emitted.

Summary table
(Note that 'c' = the speed of light = 3×10^8 m s^{-1})

Name	Details	Charge	Mass	Velocity	Absorber	Ionisation of air
Alpha particles (α)	2 protons and 2 neutrons (a helium nucleus)	+2	"large" (4)	0.1 c	sheet paper or a few cm of air	much
Beta particles (β)	electron	–1	"small" (0.0005)	0.9 c	few mm of aluminium or about 25 cm of air	little
Gamma radiation (γ)	Short burst of high freq. e.m. radiation	0	0	1.0 c	few cm of lead or metres of concrete	very little

Activity

The activity of a radioactive source is the **number** of nuclei which decay **each second**. The units of activity are becquerels (Bq), (1 Bq = 1 decay per second). The formula to calculate activity is:

becquerels — Activity = $\dfrac{\text{Number of decays}}{\text{time}}$ — seconds   or $A = \dfrac{N}{t}$

Example
In a radioactive source, 2 700 of the nuclei decay in 6 s. Calculate the activity of the source.

$$\text{Activity} = \frac{N}{t} = \frac{2700}{6} = 450 \text{ Bq}$$

Quick Test 40

1. A nucleus of Actinium-228 undergoes radioactive decay by first emitting a beta particle. The daughter product then emits an alpha particle. Write a statement to show both decays.
   (You need to know these atomic numbers: Actinium (Ac) 89; Radium (Ra) 88; Thorium (Th) 90.)
2. Which of the types of nuclear radiation causes most ionisation?
3. In a radioactive source, 3 600 nuclei decay every minute. Calculate the activity of the source.

Fission, Fusion and $E = mc^2$

Fission

Fission occurs when a large nucleus (e.g. Uranium) **breaks down** into two nuclei of **smaller mass number**. This nuclear breakdown also produces a **release of energy** and (usually) **neutrons**.

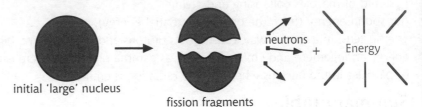

initial 'large' nucleus

fission fragments

neutrons + Energy

When the nucleus breaks up **without** any external influence causing it, it is called **spontaneous** fission.

Example

$$^{256}_{100}\text{Fm} \rightarrow {}^{140}_{54}\text{Xe} + {}^{112}_{46}\text{Pd} + 4\,{}^{1}_{0}\text{n} + \text{energy}$$

When the nucleus breaks up because it is **bombarded** with a neutron, it is called **induced** fission.

Example

$$^{1}_{0}\text{n} + {}^{235}_{92}\text{U} \rightarrow {}^{141}_{56}\text{Ba} + {}^{92}_{36}\text{Kr} + 3\,{}^{1}_{0}\text{n} + \text{energy}$$

Here the addition of one extra neutron to Uranium-235 makes the nucleus unstable and causes it to break apart into barium and krypton as well as releasing three neutrons and a great deal of energy.

Energy from fission

The reason why nuclear fission releases energy is not immediately clear – it requires modification of the concept that 'energy is conserved'. Einstein realised that in special circumstances it is possible for mass to 'disappear' and change into energy. His idea that 'mass-energy is conserved' is now completely accepted by physicists. The reason why energy is released by fission is because the total mass of the fission products is less than the total mass before fission. This **lost mass** has been **converted into energy** according to Einstein's famous equation,

$$E = mc^2$$

where, E = the quantity of released energy (in joules)

m = the loss in mass during the fission reaction (in kilograms)

c = the speed of light (= 3.0×10^8 m s^{-1})

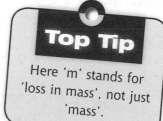

Top Tip

Here 'm' stands for 'loss in mass', not just 'mass'.

Example

Uranium can be induced to undergo fission into molybdenum and xenon as shown by this statement.

$$^{1}_{0}\text{n} + {}^{235}_{92}\text{U} \rightarrow {}^{98}_{42}\text{Mo} + {}^{136}_{54}\text{Xe} + 2\,{}^{1}_{0}\text{n} + \text{energy}$$

Use the information given to calculate the energy released.

particle	mass (kg)
$^{98}_{42}$Mo	1.62486×10^{-25}
$^{136}_{54}$Xe	2.25557×10^{-25}
$^{235}_{92}$U	3.90089×10^{-25}
$^{1}_{0}$n	1.674×10^{-27}

Answer

Mass before fission = $1.674 \times 10^{-27} + 3.90089 \times 10^{-25}$
$= 3.91763 \times 10^{-25}$ kg

Mass after fission = $1.62486 \times 10^{-25} + 2.25557 \times 10^{-25}$
$+ 2(1.674 \times 10^{-27})$
$= 3.91391 \times 10^{-25}$ kg

Loss in mass during fission = $3.91763 \times 10^{-25} - 3.91391 \times 10^{-25} = 3.72 \times 10^{-28}$ kg

Energy released = $mc^2 = 3.72 \times 10^{-28} \times (3 \times 10^8)^2 = 3.348 \times 10^{-11}$ J

The loss in mass is so small that calculations have to be done to many significant figures. You must not round numbers when finding the loss in mass.

Fusion

Nuclear fusion occurs when two **small nuclei join** (or fuse) together and form a larger nucleus.

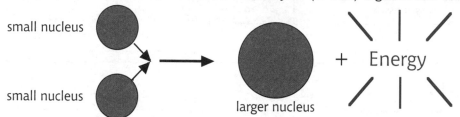

small nucleus

small nucleus

larger nucleus

+ Energy

Like fission, there is a **loss of mass** during the fusion process resulting in the 'creation' of energy according to Einstein's equation $E = mc^2$.

Example

A deuterium atom is an isotope of hydrogen which has one neutron and one proton in its nucleus. Two deuterium nuclei can be made to fuse and form a nucleus of helium according to the following statement

$$^{2}_{1}H + {}^{2}_{1}H \rightarrow {}^{3}_{2}He + {}^{1}_{0}n + \text{energy}$$

Use the data to find the energy released for one fusion reaction.

Particle	Mass (kg)
$^{2}_{1}H$	3.34258×10^{-27}
$^{3}_{2}He$	5.00473×10^{-27}
$^{1}_{0}n$	1.67444×10^{-27}

Answer

Mass before fusion = $2 \times 3.34258 \times 10^{-27} = 6.68516 \times 10^{-27}$ kg

Mass after fusion = $5.00473 \times 10^{-27} + 1.67444 \times 10^{-27} = 6.67917 \times 10^{-27}$ kg

Loss in mass during fusion = $6.68516 \times 10^{-27} - 6.67917 \times 10^{-27}$

$= 5.99 \times 10^{-30}$ kg

Energy released = $mc^2 = 5.99 \times 10^{-30} \times (3 \times 10^8)^2 = 5.39 \times 10^{-13}$ J

- Nuclear fusion must not be confused with nuclear fission although both processes result in large amounts of energy being released.
- The word 'fussion' does not exist – using it in an examination answer will lose marks because the examiner won't know if you mean 'fission' or 'fusion'.
- When describing fusion in an examination answer you must talk in terms of **nuclei** joining together. Answers using 'atoms', 'particles', etc. lose marks.

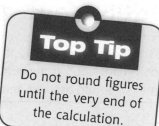

Top Tip

Do not round figures until the very end of the calculation.

Quick Test 41

1. Describe any similarities and differences between spontaneous fission and induced fission.
2. Describe any similarities and differences between nuclear fission and nuclear fusion.
3. Calculate the amount of energy released when one gram of mass is converted totally into energy.
4. A nuclear fission reaction is represented by the following statement:

$$^{235}_{92}U + {}^{1}_{0}n \rightarrow {}^{137}_{55}Cs + {}^{95}_{37}Rb + 4\,{}^{1}_{0}n$$

   (a) What type of nuclear reaction is this?
   (b) Calculate the energy released in the reaction.

	Mass/kg
$^{235}_{92}U$	390.219×10^{-27}
$^{137}_{55}Cs$	227.292×10^{-27}
$^{95}_{37}Rb$	157.562×10^{-27}
$^{1}_{0}n$	1.675×10^{-27}

Dosimetry and Safety

Absorbed dose, D

Absorbed dose, D, is a quantity which takes into account the amount of radiation **energy** absorbed and the **mass** of the tissue which absorbs it.

joules per kilogram or grays (Gy)

$$\text{absorbed dose} = \frac{\text{energy of the radiation (joules)}}{\text{mass of tissue (kilograms)}} \quad \text{or} \quad D = \frac{E}{m}$$

Example
5.0 kg of tissue absorbs 0.80 mJ of energy. Calculate the absorbed dose.

Answer
D = E/m = 0.80/5.0 = 0.16 mGy

Equivalent dose, H

Absorbed dose does not take into account the fact that **different types** of radiation can do **different** amounts of **damage** to body tissue. To allow for this, each type of radiation is allocated **a radiation weighting factor, w_R**. The larger the value of the radiation weighting factor, the more damaging the radiation, i.e. w_R is a measure of the biological effect of the radiation.

$$\text{equivalent dose} = \text{absorbed dose} \times \text{radiation weighting factor} \quad \text{or} \quad H = D \times w_R$$

The units of equivalent dose are **sieverts (Sv)**.

Top Tip
w_R has no units.

Example
A source of alpha radiation causes 2.5 kg of tissue to receive 5.30×10^{-5} J of energy. Calculate the equivalent dose. (The radiation weighting factor for alpha radiation = 20.)
D = E/m = $5.30 \times 10^{-5}/2.5$ = 21.2 μGy
H = Dw_R = 21.2 × 20 = 424 μSv

Equivalent dose rate, $\dot{H}$

It should be obvious that, if someone receives the above dose of radiation (424 μSv) over a period of 20 years, it is not as dangerous as it would be if the same dose were received over a few minutes. Equivalent dose **rate** takes this time factor into account and so gives better information about whether a person has received a dangerous dose.

$$\text{equivalent dose rate} = \frac{\text{equivalent dose}}{\text{time}} \quad \text{or} \quad \dot{H} = \frac{H}{t}$$

In the equation for equivalent dose rate, the time is often not in seconds – more often hours, days, weeks, months and years are used. This means that the corresponding units of equivalent dose rate are sieverts per hour, Sv per day, Sv per week, Sv per month or Sv per year.

Example
A technician receives 6.0 μGy of neutron radiation in a time of four hours. Calculate the equivalent dose rate. (The radiation weighting factor for neutrons is 10.)

Answer
equivalent dose = H = D × w_R = 6.0 × 10 = 60 μSv

$\dot{H} = \frac{H}{t} = \frac{60}{4}$ = 15 μSv per hour

How to reduce the equivalent dose rate

(a) Increase the distance from the source, e.g. handling the radioactive source with long tongs.
(b) Use shielding to absorb some of the energy of the radiation, e.g. a vet wears a special lead apron while using a radioactive source during an operation on an animal.

Biological harm

The factors affecting the risk of biological harm from a dose of radiation are:
- the size of the absorbed dose
- the kind of radiation received (whether the dose is from alpha, beta, gamma or slow neutrons) and
- the body organ (or tissue) receiving the dose.

We are all constantly receiving radiation from our surroundings – this radiation is outwith our control and is called **background radiation**. It is due to:
- radon gas seeping into our houses from the ground;
- radiation used in medicine (e.g. X-rays and γ-rays for medical investigation and treatment);
- gamma rays from the ground and the materials in our buildings;
- natural radiation from our food and drink (e.g. strontium 90 in milk);
- cosmic radiation from outer space (which increases the dose you receive when you fly);
- other sources caused by people (e.g. nuclear accidents).

A normal member of the public in Britain receives an **annual** effective equivalent dose of **about 2 mSv** due to background radiation but it is higher in some areas (e.g. in Aberdeen and Cornwall, due to the natural radioactivity in the granite). The annual effective equivalent dose limit is set at a **higher value for certain occupations**, e.g. for workers in the nuclear power industry.

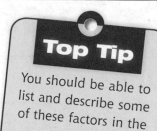

Top Tip

You should be able to list and describe some of these factors in the exam.

Half-value thickness

An **absorbing material** placed between a radioactive source and a detector causes **the count rate to decrease**. For example, the intensity of a beam of gamma radiation can be reduced using aluminium.

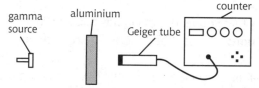

The **greater** the thickness of the aluminium, the **lower** the count rate. The graph of the results has a special shape – the count rate is **halved** for **equal increases** in the **thickness**.

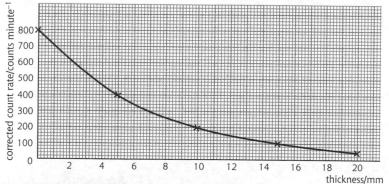

Example
In this example, every 5.0 mm the count rate is halved (800 to 400 or 400 to 200 or 200 to 100 etc). The half-value thickness of aluminium for this energy of gamma radiation is therefore 5.0 mm.

Quick Test 42

1. A sample of tissue receives an absorbed dose of 3.6 μGy. The radiation weighting factor is 20. Calculate the equivalent dose.
2. The half value thickness of lead for gamma radiation from a source is 15 mm. What thickness of lead is needed to reduce the equivalent dose rate from the source from 80 μSv per hour to 10 μSv per hour?

Units, Prefixes and Scientific Notation

Units

In Higher Physics, we use the SI system of units. This stands for 'Système International' and (obviously!) means 'International System'. This is a modern metric system used widely across the world. Using the correct units for quantities is a very important part of **good communication** between scientists. It is also important in the examination as **unit errors** in your answers will result in the **loss of marks**. The appropriate units for quantities used in the Higher Physics course are given throughout this book. Take a look at the table below for some examples.

Physical Quantity	Symbol	Unit	Unit Abbreviation
distance	s or d	metre	m
displacement	s	metre	m
speed, velocity	v	metre per second	m s^{-1}
time	t	second	s
change of velocity	Δv	metre per second	m s^{-1}
electric charge	Q	coulomb	C
electric current	I	ampere	A
voltage, potential difference	V	volt	V
period	T	second	s
frequency	f	hertz	Hz
wavelength	λ	metre	m

These quantities, and many more, should now be very familiar to you.

There is not enough room here to tabulate them all. However, a Top Tip is that you read through this book again, making your own list of all the **quantities** along with their **symbols** and **units**. You can then check the accuracy of your list by comparing it with the one provided by the SQA on their website.

The SQA gives this information in a pdf document called *Physical Quantities, Symbols and Units* (www.sqa.org.uk/ files_ccc/quant_symb_2004_final.pdf).

Prefixes

A prefix is the **extra part** added to the start of an SI unit which indicates a **multiplication factor** to be included in calculations. For example, the prefix 'milli' is shortened to 'm' and means 'thousandths' (or '$\times 10^{-3}$'). In calculating resistance, when the current is given as 6.4 mA, it is essential that the value substituted in the equation is 6.4×10^{-3} A (or 0.0064 A). The prefixes are **not given** in the Physics data booklet or in the examination paper – you need to **memorise** them.

The prefixes you need to know for the Higher Physics examination are:

Prefix name	Prefix symbol	Power of ten
pico	p	$\times 10^{-12}$
nano	n	$\times 10^{-9}$
micro	μ	$\times 10^{-6}$
milli	m	$\times 10^{-3}$
kilo	k	$\times 10^{3}$
Mega	M	$\times 10^{6}$
Giga	G	$\times 10^{9}$

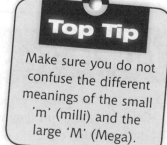

Top Tip

Make sure you do not confuse the different meanings of the small 'm' (milli) and the large 'M' (Mega).

Scientific notation

Giving a number in scientific notation means that there is only **one** figure before the decimal point. For example, the number 340 can also be written as 34×10^1 OR 3.4×10^2. All three forms have exactly the same value, but only 3.4×10^2 states the value in scientific notation. One advantage of using scientific notation is that very large (or very small) values can be written in a short form and can be entered into a calculator.

For example, the speed of light can be stated as 'three hundred million metres per second' or '300 000 000 m s^{-1}' or '300×10^6 m s^{-1}' or "3.0×10^8 m s^{-1}".

(Only this last statement is the value of the speed of light in scientific notation.)

In calculators this number is entered by using the following five keystrokes.

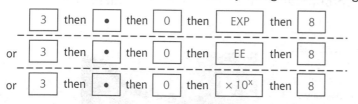

| 3 | then | • | then | 0 | then | EXP | then | 8 |

or | 3 | then | • | then | 0 | then | EE | then | 8 |

or | 3 | then | • | then | 0 | then | × 10^x | then | 8 |

Another advantage of using scientific notation is that the number of **significant figures** can easily be seen. This is important because it conveys to any reader **how accurately** you are quoting a value. For example, the charge on an electron can be quoted as:

This statement gives the charge on the electron to two significant figures. It implies that the value lies anywhere between 1.55×10^{-19} C and 1.64×10^{-19} C. → "1.6×10^{-19} C"

OR as

This statement gives the charge on the electron to four significant figures. It implies that the value lies anywhere between 1.6015×10^{-19} C and 1.6024×10^{-19} C. → "1.602×10^{-19} C"

The second statement implies **much greater precision**.

Counting significant figures

Start counting (from the left-hand side) at the first **non-zero figure** – stop counting at the last figure.

Examples

(a) 6.63×10^{-34}

start here, one two three

This number is being quoted to three significant figures.

(b) 238.02

one two three four five

This number is being quoted to five significant figures (and is <u>not</u> given in scientific notation).

Top Tip

Do not round the answer to the first part of a two-part calculation – this can make the final answer inaccurate.

Do not confuse giving an answer to a certain number of 'decimal places' with giving an answer to a number of 'significant figures'. In example (b) the number is given to two decimal places but to five significant figures. You are asked in the examination to give answers 'to an appropriate number of significant figures'. To do this, round the final answer to the same number of significant figures as the data in the question.

Correct rounding to an appropriate number of significant figures

Example

Suppose the data in a particular question is given to three significant figures. Your calculator displays the answer to the calculation as $4.738709677 \times 10^{14}$. The third significant figure in this answer is the '3'. However, because the fourth figure is '8', the '3' is rounded up and the answer should be written as 4.74×10^{14}. Do not round up when the next figure is less than '5'.

Uncertainties

When a measurement is made of any physical quantity (distance, time, mass, temperature, voltage etc) there is **imprecision** in the result. This imprecision is usually called **experimental uncertainty**. Experimental uncertainties arise for a number of different reasons such as the **limitations of the apparatus** being used or issues about the way the experimenter **uses the apparatus**. A good example of the limitations of a piece of apparatus is when a student uses a ruler to measure length. The smallest division on the ruler's scale is one millimetre, meaning that the length cannot be measured more accurately than to about half a millimetre. This is a **scale-reading uncertainty**. Repeated measurements are likely to produce a series of results which vary equally above and below the 'correct' result – this is called a **random uncertainty**.

An example of an experimenter **using apparatus inappropriately** is when the surface of a liquid in a measuring cylinder is not viewed correctly.

Taking a reading incorrectly.

When viewed from a high angle the experimenter sees the surface of the liquid at a different (and wrong!) position relative to the scale on the measuring cylinder.

The reading should be taken by looking along a horizontal line, level with the surface of the liquid.

Taking a reading correctly.

Incorrect use or set-up of apparatus can produce a series of results which are, on average, higher than (or lower than) the 'correct' result – this is called a **systematic uncertainty**.

Showing uncertainties

When the result of an experimental measurement is recorded, the uncertainty about its accuracy should be shown as well as the result itself. For example, a student plans to drop a ball from a height. She measures the height as 2.50 metres. She estimates that the uncertainty in this measurement is ±2 centimetres (±0.02 m) – this means that she believes the value of the height lies between 2.48 and 2.52 metres.
She should record her result as 2.50 ± 0.02 m.
The **absolute uncertainty** in her measurement is the value of the imprecision, i.e. ±0.02 m.
The **fractional uncertainty** in her measurement is the absolute uncertainty divided by the value of the measurement, i.e. 0.02/2.50 = 0.008.
The **percentage uncertainty** in her measurement is the fractional uncertainty multiplied by 100, i.e. 0.008 × 100 = 0.8 %.
When a measurement is repeated, a series of results is produced.

The **mean value** is then calculated from
$$\text{mean} = \frac{\text{sum of the results}}{\text{number of results}}$$

The mean is the **best estimate** of a 'true' value for the measurement.
The approximate random uncertainty in the mean is calculated from:

$$\text{approximate random uncertainty} = \frac{\text{maximum value} - \text{minimum value}}{\text{number of measurements taken}}$$

Top Tip

This formula is listed in the Physics data booklet.

Increasing the number of measurements taken reduces the approximate random uncertainty (because the range is being divided by a larger number).

Example

An object is repeatedly dropped from a height of 2.5 m and measurements made of the time of fall.
The results are; 0.71 s, 0.68 s, 0.70 s, 0.73 s and 0.72 s
(a) Calculate the mean time of fall.
(b) Calculate the approximate uncertainty in the mean time of fall.
(c) Make a statement giving the final result of the experiment.

Don't forget to give units after the mean and the uncertainty.

Answer

(a) Mean = sum/number of measurements = (0.71 + 0.68 + 0.70 + 0.73
$\qquad$ + 0.72)/5 = 0.708 = 0.71 s.
(b) Random uncert. = (highest – lowest)/number = (0.73 – 0.68)/5 = 0.05/5 = 0.01 s.
(c) Time of fall = 0.71 ± 0.01 s.

Combining uncertainties

In many experiments, two (or more) physical quantities are measured and the results combined in a calculation. An example of this is when current and voltage are measured in order to calculate resistance. To find a good estimate of the percentage uncertainty in the final result, use the larger (or largest) of the percentage uncertainties in the measurements.

Example

A student takes readings from the meters in the following experiment. Calculate the resistance of the resistor and give the uncertainty in its value.

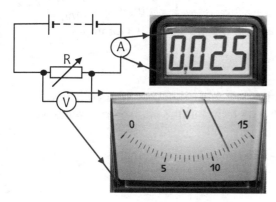

Answer

(a) R = V/I = 12/0.025 = 480 Ω.
(b) The final digit on the display of the ammeter can vary by
$\qquad$ ±1. This makes the uncertainty in the scale reading of the
$\qquad$ ammeter ±0.001 A.
The fractional uncertainty is 0.001/0.025 and
the percentage uncertainty is 4%.
Half a scale division on the display of the voltmeter is
0.25 V. This makes the fractional uncertainty 0.25/12
and the percentage uncertainty is 2%.
The uncertainty in the resistance is therefore ±4%.
4% of 480 = 19, which rounds to 20.
The resistance is therefore 480 ± 20 Ω.

Quick Test 43

1. What are the units of each of the quantities in the table?
2. Change 56 pA into amperes.
3. Change 480 nm into metres.
4. Change the following numbers into scientific notation:
    (a) 340 m s⁻¹
    (b) 2500 kV
    (c) 4700 µF
5. A student uses an old, wooden metre stick to measure the height of a bench. The ends of the metre stick are worn away. What type of uncertainty does this introduce into the measurements?

Quantity	Unit
mass	
weight	
electric charge	
capacitance	
period	
absorbed dose	

Unit tests and Practical Assessment

Unit tests

Students must pass one test for **each** of the three units of the course (Mechanics and Properties of Matter, Electricity and Electronics, Radiation and Matter).

This formal assessment examines both Knowledge & Understanding and Problem Solving. Each assessment (test) lasts for 45 minutes and consists of a total of 30 marks.

The threshold of attainment (the **pass mark**) is **18 marks** (i.e. 60% of the total). The SQA does not grant a course award to any student who has not achieved a pass for each unit. However, more than one attempt to pass a test is permitted. It is wise to **be as well prepared as possible** before attempting a test. Look at the list below for good preparation guidelines.

- Study regularly over a long period of time – not just a day or two before a test.
- Use notes and textbooks to understand and learn the work as well as possible.
- Write facts, symbols and diagrams from memory and then make a comparison with notes and textbooks to identify gaps/weaknesses in your knowledge and understanding.
- Write out definitions of all the quantities in the relationships listed in the Physics data booklet.
- Memorise the units for all the quantities.
- Review and learn from mistakes made in homework tasks.
- Make sure that you can use every formula to calculate each of the quantities given within it.
- Scrutinise the content statements for each unit to make sure you know what is expected of you. Teachers often provide lists of these content statements, but they can also be found in the course arrangements document on the SQA website.
- Write out explanations of facts, derivations and definitions listed in the content statements and practise them frequently.

Top Tip

Studying regularly is more effective than last minute 'cramming'.

Practical assessment

During the course, students must have been actively involved in at least **one experiment** and have completed a **write-up** to **certain standards** laid down by the SQA. You should be given guidance about the required standards and an **opportunity to redraft** the write-up if it did not initially meet those standards. The SQA only grants a course award to students who have achieved a successful write-up.

Advice for achieving a successful write-up

- Write in what is known as the 'impersonal', 'passive' voice in the past tense. This means that you should try to write like this: 'the light gate was connected to a computer', not like this 'I connected the light gate to a computer'. This is not an essential requirement, but it is common practice in scientific reports.
- Sentences should be brief, clear and to the point.

- Lengthy descriptions are not required - a few short sentences are sufficient.
- It is useful to structure your report under certain **headings** to avoid missing out important sections. The following headings are suggested as well as more detail about how to meet the required standards.

Title

This can be copied from the instruction sheet.

Aim or objective

A brief statement of the purpose of the experiment

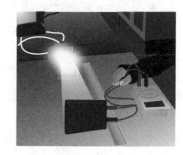

Apparatus

- Write a list of all the apparatus used in the experiment.
- Draw a simple line diagram of the experimental arrangement.
- Label each piece of equipment shown in the diagram.

Procedures

- Describe **how** all the measurements were taken or observations made.
- Describe **how** the independent variable was altered. (The independent variable is the factor which the experimenter is changing/controlling.)

Readings/observations

- Draw a **table of results** with suitable headings. Remember to give quantities and their units. (Arrange the values of the independent variable in ascending order or descending order.)

Analysis

- You may need to use data from the table of results to calculate other quantities (for example, the square or inverse of one of the measurements). If so, these new values should be given in another table or added to the first table.
- The results/calculations should be used to plot a **line graph**. Make sure that the **axes** are clearly **labelled**, giving the names of the **quantities** and their **units**. (It is usual to plot the independent variable along the horizontal axis.) Make sure that all the points are **correctly plotted**.
- Your graph must show a **straight line** or **curve** of **'best fit'**. The line must not join your results 'dot-to-dot'. Rather, it should show the 'average' trend of the results. The best fit line will pass through (or very close to) most of the points, but it is also likely that some points will be above and some below the line. Remember that only a **straight**, **diagonal line through the origin proves** that the quantities on the 'x' and 'y' axes are directly proportional to each other.

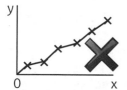 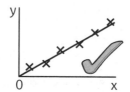 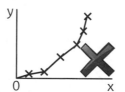 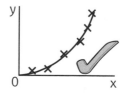

Uncertainties

The detail required here depends on the particular experiment.
However, at least one the following should be included.

- Uncertainties in individual readings should be quoted as '±imprecision'.
- Calculations of the **mean value** and the **approximate random uncertainty** in the mean should be given where appropriate (i.e. only when the **same measurement is repeated** several times).
- Uncertainties could be expressed as their absolute values or given as a **percentage**.

Top Tip

Be certain about the values of your uncertainties!

Conclusion

The **conclusion should link back to the aim** of the experiment, stated at the start of the write-up. Statements should be made about as many of the following areas as are relevant.

- If there is an overall pattern to the observations, what is that pattern?
- If there is an overall trend in the analysed information, what is that trend?
- What relationship between the variables is proved by the results?
- Do the results produce a final answer for the value of a physical quantity?
  (for example, the acceleration due to gravity.)

Evaluation

The detail required here depends on the particular experiment. However, the evaluation is likely to contain, as appropriate, a few sentences commenting on at least **some of the following**:

- The **effectiveness** of the experimental procedures. Did the results produce a final value equal to an accepted value (within the limits of the experimental uncertainties)? Did the results produce a convincing conclusion about a relationship between variables?

- Suggestions of possible ways of **making improvements** to the experiment – procedure and/or apparatus. Do not write vague comments such as 'use a more accurate meter'. Be more precise, such as 'replace the voltmeter, which had a reading uncertainty of ± 0.05 V, with a digital meter which can measure to ± 0.01 V'. This would reduce the percentage uncertainty in the final calculated value of resistance.'

- How the procedures made sure that **variables** were **controlled**.

- Discussion of the sources of the **uncertainties** in the experiment. There could be scope here to **compare percentage uncertainties** and so conclude what change would best improve the experiment.

- Describe any **limitations** with the **equipment**. Discuss any other factors which might be changed to improve the experiment. (For example, an initial experiment might have been designed to measure the time of fall of an object dropped through heights of up to 1.0 m. This could be redesigned to use heights up to 2.0 m to give an improved range and number of results.)

Answers to Quick Tests

Unit 1: Mechanics and Properties of Matter

Page 5: Scalars and Vectors
Quick Test 1

1. A scalar quantity has only size (or magnitude). A vector quantity needs a direction as well as a magnitude.
2. 'Distance' and 'speed' are scalars. 'Displacement' and 'velocity' are vectors.
3. 335°.
4. 10 m s^{-1}
5. 2.5 km h^{-1} bearing 108° (or 0.69 m s^{-1} bearing 108°).

Page 7: Adding together Vector Quantities
Quick Test 2

1. 33 N to the left
2. 11 N to the right
3. 250 m bearing 127°
4. 13 m s^{-1} bearing 337°
5. 38.7 N bearing 065°

Page 9: Resolving Vectors into Components
Quick Test 3

1. 67.6 N to the right
2. horizontal = 47 m s^{-1}, vertical = 34 m s^{-1}
3. 4890 N
4. No (because the maximum value of friction is greater than the component of weight (113 N)).
5. 370 N (driving force + component of weight – friction).

Page 11: Acceleration
Quick Test 4

1. False. (Its velocity will be increasing very quickly, but it may have only just started moving.)
2. True. (Acceleration only states how quickly velocity is changing, not the value of the velocity.)
3. The object's velocity is increasing by 4.0 m s^{-1} each second.

4. $a = (v - u)/t = (11.7 - 2.6)/6.5$
   $= 9.1/6.5 = 1.4 \text{ m s}^{-2}$
5. $a = (v - u)/t = (21.7 - 37.0)/3.4$
   $= -15.3/3.4 = -4.5 \text{ m s}^{-2}$.
   (N.B. your answer must be negative.)
6. 29.4 m s^{-1} ($= 3.0 \times 9.8$)

Page 13: Graphs of Motion
Quick Test 5

1. B, C & F
2. (a) $- 2.0 \text{ m s}^{-2}$ (b) $+ 60 \text{ m}$

Page 15: The Equations of Motion
Quick Test 6

1. 100 m
2. 1.0 m s^{-2}
3. (a) 7.8 s (b) 77 m s^{-1}
4. 3.3 s
5. 748 m

Page 17: Projectiles
Quick Test 7

1. (a) 3.67 s (b) 66 m
2. (a) 42.4 m s^{-1} (b) 24.5 m s^{-1} (c) 2.5 s
   (d) 30.6 m   (e) 212 m
3. (a) same (b) same (c) less (d) less (e) less

Page 19: Newton's Laws of Motion
Quick Test 8

1. An upward force of 50 N
2. The car experiences a negative acceleration and slows down.
3. (a) 4890 N (b) 7990 N (8000 N)
   (c) 3.3 m s^{-2}
4. 7.35 N (equal to the weight of the book).

Page 21: Work, Energy and Power
Quick Test 9

1. Work is in joules and is a scalar. Energy is in joules and is a scalar. Power is in watts and is a scalar.
2. $E_w = F d = 96 \times 3.5 = 336 \text{ J}$
3. $E_k = \frac{1}{2}mv^2 = \frac{1}{2} \times 12\,500 \times 32^2 = 6.4 \times 10^6 \text{ J}$
4. Initial $E_k = \frac{1}{2}mv^2 = \frac{1}{2} \times 1400 \times 24^2$
   $= 403\,200 \text{ J}$.
   Final $E_k = \frac{1}{2}mv^2 = \frac{1}{2} \times 1400 \times 8^2$
   $= 44\,800 \text{ J}$. Loss in $E_k = 3.6 \times 10^5 \text{ J}$
5. $v = \sqrt{(2\,g\,h)} = \sqrt{(2 \times 9.8 \times 15)} = 17.1 \text{ m s}^{-1}$
6. $P = E/t = 8400/120 = 70 \text{ W}$

Answers to Quick Tests

Page 23: Collisions and Momentum
Quick Test 10

1. (a) momentum = mass × velocity (b) $p = m v$
2. kilogram metres per second (kg m s^{-1})
3. Vector
4. 1.5 m s^{-1} in the same direction as the 6.0 kg object had been travelling.
5. (a) Total momentum before = total momentum after
   $m_1 u_1 + m_2 u_2 = m_1 v_1 + m_2 v_2$
   $\Rightarrow (6.0 \times 0) + (8.0 \times 0)$
   $= (8.0 \times 4.0) + 6v$
   $\Rightarrow 0 = 32 + 6v$
   $\Rightarrow 6v = -32$
   $\Rightarrow v = -5.3$ m s^{-1} i.e. the 6.0 kg object moves at a speed of 5.3 m s^{-1} in the opposite direction to the 8.0 kg.
   (b) The total E_k is initially zero. The E_k increases (from the energy initially stored in the spring). It is not elastic.

Page 25: Impulse
Quick Test 11

1. (a) impulse = force × time
   (b) newton second (N s) (c) vector
2. impulse = F t = 16 × 52 = 832 N s
   (note that the value of the mass is irrelevant.)
3. change in momentum = impulse = F t
   = 75 × 18 = 1350 kg m s^{-1} (note that the value of the mass is irrelevant.)
4. change in momentum = impulse $\Rightarrow$
   (m v – m u) = F t $\Rightarrow$ (4.5 × v – 4.5 × 1.3)
   = 36 $\Rightarrow$ v = 9.3 m s^{-1}
5. change in momentum = impulse = area under the force/time graph = $\frac{1}{2}$ × 86 × 10^{-3} × 220
   = 9.46 kg m s^{-1}

Page 27: Density
Quick Test 12

1. (a) density = mass/volume (b) kilograms per cubic metre (kg m^{-3}) (c) scalar
2. ρ = m/V = 5.60/0.0020 = 2 800 kg m^{-3}
3. m = ρ V = 19 300 × 1.3 × 10^{-6}
   = 0.02509 = 0.0251 kg (= 25.1 g)
4. density = mass/volume. The mass remains the same. The volume is approximately the same. The density is (approx.) the same.
5. density = mass/volume. The mass remains the same. The volume is much larger (approximately 1000 times greater). The density is much smaller (approximately 1000 times smaller).

Page 29: Pressure
Quick Test 13

1. P = F/A = 690/0.30 = 2300 N m^{-2}
   (OR 2300 Pa)
2. Area = 1.5 cm^2 = 1.5 × 10^{-4} m^2. P = F/A
   = 750/1.5 × 10^{-4} = 5.0 × 10^6 N m^{-2}
   (OR 5.0 × 10^6 Pa)
3. The molecules of a gas are in constant, random motion and keep colliding with the walls of the container. These collisions cause a force (and so a pressure) on the walls.
4. The density of the liquid. The gravitational field strength of the planet. The depth below the surface of the liquid.
5. P = ρ g h = 1.02 × 10^3 × 9.8 × 56
   = 559 776 = 560 000 Pa

Page 31: Upthrust and Flotation
Quick Test 14

1. Unit of upthrust are newtons. Upthrust is a vector quantity.
2. (a)

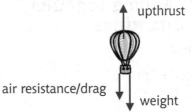

air resistance/drag ... upthrust ... weight

   (b) upthrust = weight + drag
   (because constant speed means the forces are balanced)
3. upthrust = weight of ship = m g
   = 3 600 000 × 9.8 = 35 280 000
   = 35 000 000 N
4. weight = m g = 2 400 000 × 9.8
   = 23 520 000 N.
   F = 25 000 000 – 23 520 000
   = 1 480 000 N upwards.
   The submarine accelerates upwards at a rate of
   a = F/m = 1 480 000/2 400 000 = 0.62 m s^{-2}.

Page 33: The Gas Laws 1
Quick Test 15

1. Pressure vs $\frac{1}{\text{Volume}}$ graph

2. $P_1 V_1 = P_2 V_2$
   $\Rightarrow$ 1.0 × 10^5 × 36 = 2.5 × 10^5 × V$_2$
   $\Rightarrow V_2 = \dfrac{1.0 \times 10^5 \times 36}{2.5 \times 10^5} = 14.4$ cm^3

3. (a) 273 K (b) 300 K (c) 250 K (d) 546 K
4.

Pressure/Pa

(line graph of Pressure/Pa vs Temperature/K, straight line from origin)

Temperature/K

5. 22 °C = 295 K;   100 °C = 373 K

$$\frac{P_1}{T_1} = \frac{P_2}{T_2} \Rightarrow \frac{1.01 \times 10^5}{295} = \frac{P_2}{373}$$

$$\Rightarrow P_2 = \{373 \times 1.01 \times 10^5\}/295$$
$$= 1.28 \times 10^5 \text{ Pa}$$

Page 35: The Gas Laws 2
Quick Test 16

1. $T_1 = 25$ °C $= 298$ K
   $T_2 = 100$ °C $= 373$ K

$$\frac{V_1}{T_1} = \frac{V_2}{T_2} \Rightarrow \frac{2.5}{298} = \frac{V_2}{373}$$

$$\Rightarrow V_2 = \frac{(373 \times 2.5)}{298} = 3.13 \text{ m}^3$$

2. $$\frac{P_1 V_1}{T_1} = \frac{P_2 V_2}{T_2}$$

$$\Rightarrow \frac{1.5 \times 10^5 \times 7.2}{300} = \frac{2.5 \times 10^5 \times V_2}{400}$$

$$\Rightarrow V_2 = \frac{400 \times 1.5 \times 10^5 \times 7.2}{300 \times 2.5 \times 10^5} = 5.76 \text{ m}^3$$

3. When the temperature decreases, the kinetic energy of the molecules decreases. The molecules move slower, their collisions with the walls are less hard and less frequent. This reduces the force on the walls and the pressure decreases.

Unit 2: Electricity and Electronics

Page 37: Electric Charges and Fields
Quick Test 17

1. The balloon has lost negative electrons (to the jersey) and so is positively charged.

2. (a) Q (b) coulombs (C) (c) scalar
3. no. of electrons = total charge/electronic charge = $4.5 \times 10^{-15}/1.6 \times 10^{-19} = 30\ 000$
4. $W = Q V = 7.4 \times 10^{-14} \times 60 = 4.44 \times 10^{-12}$ J
5. $E_k = \frac{1}{2} mv^2 = QV = 1.6 \times 10^{-19} \times 400$

   $= 6.4 \times 10^{-17}$ J
   $v^2 = 2 \times 6.4 \times 10^{-17}/9.11 \times 10^{-31}$
   $\Rightarrow v = 1.19 \times 10^7$ m s^{-1}

Page 39: Circuits: Series and Parallel Connections
Quick Test 18

1. (a) 24.1 Ω (b) 3.0 Ω (c) 10 kΩ
2. $V_1 = 3.0$ V $V_2 = 6.0$ V
3. (a) $R = V/I = 3.6/12 = 0.30$ Ω
   (b) $P = I V = 3.6 \times 12 = 43.2$ W

Page 41: e.m.f. and Internal Resistance
Quick Test 19

1. Each coulomb of charge passing through the supply is given 9.0 J of energy.
2. (a) 0.1 V (b) r = lost volts/I = 0.1/0.2
   $= 0.5$ Ω
3. (a) I = e.m.f./R_{total} = 3.2/16 = 0.20 A
   (b) $V = I R = 0.20 \times 15 = 3.0$ V
   (c) $P = I V = 0.20 \times 3.0 = 0.60$ W

Page 43: The Wheatstone Bridge
Quick Test 20

1. 4
2. (a) False (the potential dividers are in parallel)
   (b) True (c) True (d) False (e) True
3. $R_1/R_2 = R_3/R_4 \Rightarrow R_{LDR}/6.8 = 56/22$
   $\Rightarrow R_{LDR} = 17.3090909 = 17.3$ kΩ

Page 45: Alternating Current and Voltage
Quick Test 21

1. (a) True (b) False (it is not just that they <u>can</u> flow both ways – they do regularly reverse their direction of flow).
   (c) True (d) False – it measures the r.m.s. value.
2. V_p = no. of division × y-gain = $4 \times 5 = 20$ volts.
3. period = no. of divisions × timebase setting
   $= 2 \times 4.0 \times 10^{-3} = 8.0 \times 10^{-3}$ s
   Frequency = 1/period = $1/8.0 \times 10^{-3}$
   $= 125$ Hz.
4. $V_p = V_{r.m.s.} \times \sqrt{2} = 230 \times \sqrt{2} = 325$ V.

Page 47: Capacitors

Quick Test 22

1. $C = Q/V = 4.95 \times 10^{-5}/1.5 = 3.3 \times 10^{-5}$ F
   $(= 33 \ \mu F)$
2. $Q = C V = 47 \times 10^{-6} \times 9.0 = 4.23 \times 10^{-4}$ C
3. $I = V/R = 6/120 = 0.05$ A (note that the value of capacitance is irrelevant).

Page 49: Capacitors in Circuits

Quick Test 23

1. $E = \frac{1}{2} Q V = 0.5 \times 6.4 \times 10^{-4} \times 6.0$
   $= 1.92 \times 10^{-3}$ J
2. $E = \frac{1}{2} C V^2 = \frac{1}{2} 430 \times 10^{-9} \times 12^2$
   $= 3.096 \times 10^{-5} = 3.1 \times 10^{-5}$ J
3. $E = \frac{1}{2} Q^2/C = \frac{1}{2} (3.4 \times 10^{-4})^2/6800 \times 10^{-6}$
   $= 8.5 \times 10^{-6}$ J
4. The a.c. current gradually increases because a capacitor allows a higher current at a higher frequency.
5. Your answer could describe the usefulness of storing charge and energy – for example in a camera's electronic flash, where the energy is suddenly released in a bright flash of light.

   Other answers could be details on blocking d.c. signals or filtering high frequencies or smoothing pulsed d.c.

Page 51: The Operational Amplifier (op-amp) 1

Quick Test 24

1. $V_o = - V_1 \times R_f/R_1 = - 0.25 \times 60/1.5 = - 10$ V
2. $V_o = - V_1 \times R_f/R_1 = - (- 0.25) \times 120/1.5$
   $= + 20$ V
   But because the op-amp saturates, the maximum output voltage will be around 13 V (for a ± 15 V supply).
3. $V_o = - V_1 \times R_f/R_1$
   $\Rightarrow 6.5 = - V_1 \times 2.5 \times 10^6/6.8 \times 10^3$
   $\Rightarrow V_1 = - 0.01768 = - 0.018$ V
4. The output is not a square wave. Although the **peak** input voltage causes saturation and flattening of the output waveform, the gain is not high enough to cause lower values of input voltage (e.g. 0.5 V) to give an output voltage of 13 V.

Page 53: The Operational Amplifier (op-amp) 2

Quick Test 25

1. $V_o = (V_2 - V_1) \times R_f/R_1 = (7.5 - 6.8) \times 20/2$
   $= 0.7 \times 10 = 7.0$ volts
2. $V_{in} = V_1 \ V_o = (V_2 - V_1) \times R_f/R_1 \Rightarrow 8.0$
   $= (1.7 - V_1) \times 100/5 \Rightarrow (1.7 - V_1) = 8/20$
   $\Rightarrow (1.7 - V_1) = 0.4 \Rightarrow V_1 = 1.7 - 0.4$
   $= 1.3$ volts

Unit 3: Radiation and Matter

Page 55: Wave Terms and Properties

Quick Test 26

1. Its amplitude increases.
2. There are three waves in 21 m.
   The wavelength is 7 m.
3. (a) each wave is produced (or passes any point) in a time of 25 m s (= 0.025 s)
   (b) $f = 1/T = 1/0.025 = 400$ Hz.
4. Interference.

Page 57: Phase, Path Difference and Interference

Quick Test 27

1. Waves combine crest with crest and trough with trough.
2. (a) Waves with greater amplitude are produced.
   (b) Constructive interference.
3. Path difference = 990 – 840 = 150 mm.
   Five complete waves can fit into this path difference, (i.e. path difference = 150 = 5 λ). The waves from the two sources therefore meet in phase at P and constructive interference occurs.
4. Second minimum is due to the path difference being equal to 1.5 λ. So 1.5 λ = 0.3 m. So λ = 0.2 m.
5. Use a longer wavelength of light (e.g. red rather than blue). Use a grating with slits closer together. Move the screen further away.

Page 59: Spectra using Grating and Prisms

Quick Test 28

1. (a) red (b) largest wavelength (c) lowest frequency.

2. (a) $n\lambda = d\sin\vartheta$
   $\Rightarrow 2 \times \lambda = 2.0 \times 10^{-6}\sin 28.7$
   $\Rightarrow \lambda = 4.8 \times 10^{-7}$ m (= 480 nm)
   (b) blue
3. (a) refraction (higher frequencies refracting more than lower frequencies) (b) blue/violet.
4. Violet, indigo, blue, green, yellow, orange, red.

Page 61: Refraction: Snell's Law
Quick Test 29

1. Frequency.
2. data: n_1 = refractive index of air = 1.00
   n_2 = refractive index of glass = 1.60
   $\vartheta_1 = 53°$ and ϑ_2 = angle to be calculated
   $n_1\sin\vartheta_1 = n_2\sin\vartheta_2$
   $\Rightarrow 1.0 \times \sin 53° = 1.60 \times \sin\vartheta_2$
   $\Rightarrow \sin\vartheta_2 = 1.0 \times \sin 53° / 1.60$
   $\Rightarrow \sin\vartheta_2 = 0.4991$
   $\Rightarrow \vartheta_2 = 29.9°$
3. (a) $n_1\sin\vartheta_1 = n_2\sin\vartheta_2$
   $\Rightarrow 1.0 \times \sin 58° = n_2\sin 28°$
   $\Rightarrow n = 1.8$
   (b) $\lambda_1/\lambda_2 = n_2/n_1 \Rightarrow 480/\lambda_2 = 1.8/1.0$
   $\Rightarrow \lambda_2 = 267$ nm
   (c) frequency in material = frequency in air
   = $v/\lambda = 3.0 \times 10^8/4.8 \times 10^{-7}$
   = 6.25×10^{14} Hz

Page 63: Total Internal Reflection and the Critical Angle
Quick Test 30

1. The critical angle is the value of the angle of incidence when the refracted ray just emerges along the surface of the medium (or the angle of refraction = 90°).
2. The angle of incidence must be greater than the critical angle.
3. $\sin\vartheta_c = 1/n = 1/1.33 \Rightarrow \vartheta_c = 48.8°$
4. $\sin\vartheta_c = 1/n = 1/2.42 \Rightarrow \vartheta_c = 24.4°$ The angle of incidence (= 26°) is greater than the critical angle – so total internal reflection does occur.
5. Reflection occurs twice in a periscope (see the diagram in the middle of page 63). Although 'left' and 'right' are swapped on the first reflection, they are swapped back again on the second reflection.

Page 65: Irradiance
Quick Test 31

1. $I = P/A = 6.0/1.5 = 4.0$ W m^{-2}.

2. power = energy/time = 280/7.0 = 40 watts
   area = length × breadth = 0.50 × 1.60
   = 0.80 m^2
   $I = P/A = 40/0.80 = 50$ W m^{-2}
3. Calculating I d^2 for each pair of readings gives

Distance (m)	0.10	0.30	0.50	0.70	0.90	1.10	1.30
Irradiance (W m^{-2})	5.0	0.56	0.20	0.10	0.062	0.041	0.030
I d^2		0.050	0.0504	0.050	0.049	0.050	0.0496	0.051

Because the values of I d^2 are constant (= 0.050), the source of light is a point source.

4.

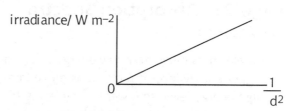

5. $I_1 d_1^2 \qquad = I_2 d_2^2$
   $0.12 \times (0.75)^2 = I_2 \times (1.5)^2$
   $0.0675 \qquad = I_2 \times 2.25$
   $I_2 \qquad = 0.030$ W m^{-2}
   This can also be answered by saying that because the distance has been doubled, (and I α 1/d^2), the irradiance must be a quarter of its initial value.

Page 67: The Photoelectric Effect 1
Quick Test 32

1. $E = hf = 6.63 \times 10^{-34} \times 5.6 \times 10^{14}$
   $= 3.7128 \times 10^{-19} = 3.7 \times 10^{-19}$ J
2. $E = hf = hv/\lambda$
   $= 6.63 \times 10^{-34} \times 3.0 \times 10^8/540 \times 10^{-9}$
   $= 3.683333 \times 10^{-19}$
   $= 3.7 \times 10^{-19}$ J
3. $E = hf = 6.63 \times 10^{-34} \times 7.9 \times 10^{14}$
   $= 5.2377 \times 10^{-19}$ J
   Max E_k = photon energy – work function
   $= hf - hf_o = 5.2377 \times 10^{-19} - 3.4 \times 10^{-19}$
   $= 1.8377 \times 10^{-19} = 1.8 \times 10^{-19}$ J

Page 69: The Photoelectric Effect 2
Quick Test 33

1. More photons per second are incident on the surface.
2. More photons per second mean more total energy per second. This means the power is increased. Irradiance = power/area, so irradiance is greater.
3. High irradiance means many photons per second but if the energy of each photon is

below the work function, the electrons do not receive enough energy to escape the attractive forces of the metal.

4. Each photon still has the same energy as before (there are just more of them per second). Each electron therefore receives the same energy as before.

Page 71: Atomic Energy Levels and Emission Spectra

The Quick Test for these pages is on page 73, questions 1 to 4.

Page 73: Absorption Spectra

Quick Test 34

1. (a) electrons move from low energy levels to higher energy levels (b) electrons move from high energy levels to lower energy levels with the emission of a photon which has energy equal to the gap between the levels.

2. (a) the ground state; (b) excited states (c) the ionisation level.

3. (a) energy required = gap between E_1 and E_4
   $= 5.55 \times 10^{-19} - 0.96 \times 10^{-19}$
   $= 4.59 \times 10^{-19}$ J
   (b) energy released = gap between E_4 and E_2
   $= 2.45 \times 10^{-19} - 0.96 \times 10^{-19}$
   $= 1.49 \times 10^{-19}$ J
   (c) E = hf. Highest frequency is due to a transition through the largest gap, i.e. E_4 to E_1 which from (a) = 4.59×10^{-19} J.
   so f = E/h = $4.59 \times 10^{-19}/6.63 \times 10^{-34}$
   $= 6.92 \times 10^{14}$ Hz
   (d) Longest wavelength is equivalent to the lowest frequency, which is due to the smallest gap, i.e. E_4 to $E_3 = 4.4 \times 10^{-20}$ J
   so f = E/h = $4.4 \times 10^{-20}/6.63 \times 10^{-34}$
   $= 6.64 \times 10^{13}$ Hz. Wavelength, λ = v/f
   $= 3.0 \times 10^8/6.64 \times 10^{14}$ $= 4.52 \times 10^{-6}$ m

4. A particular line is due to electrons making a particular transition between energy levels. More electrons are making that particular transition per second. This produces more energy at that particular frequency and hence a brighter line.

5. The atoms are taking in radiated energy and its electrons are moving up from lower energy levels to higher levels.

6. The emission spectrum shows the colours which are missing (from the visible spectrum) in the absorption spectrum.

7. Some of the light emitted from the Sun is absorbed by cooler atoms in the outer layers of its atmosphere.

Page 75: The Laser

Quick Test 35

1. They are similar in that they are both due to an electron falling from a higher energy level to a lower level and energy being emitted as a photon. They are different in that spontaneous emission occurs without being caused by any influence from outside the atom whereas stimulated emission results from sending a photon (with energy equal to the energy gap) into the atom.

2. Both are random processes – the time when a particular atom will emit light (or decay) cannot be predicted.

3. Stimulated emission of radiation.

4. They have the same frequency, energy, phase and direction.

5. The power is concentrated on to a very small area. This causes a very high value of irradiance which could damage the delicate retina of the eye.

Page 77: Conductors, Insulators and Semiconductors

Quick Test 36

1. (a) copper, silver, gold, carbon
   (b) wood, paper, plastic, rubber, glass, diamond
   (c) silicon, germanium.

2. Three.

Page 79: The p-n Junction Diode

Quick Test 37

1. It is a very narrow region either side of the junction where there is a lack of free charge carriers.

2. The p-side must be connected to the positive terminal of the supply, i.e.

3. In the forward biased diode, electrons combine with holes at the junction. There is one photon of light for each combination of an electron with a hole.

Page 81: Using the Photodiode and the MOSFET

Quick Test 38

1. Photovoltaic and photoconductive.

2. Photoconductive. (No external power supply is used in photovoltaic mode.)

3. Photoconductive.
4.

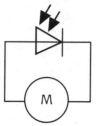

5. It is made up of free negative electrons ('n' for negative).
6.

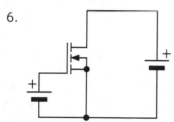

Page 83: Atomic Structure and Nomenclature

Quick Test 39

1. The fact that most of the alpha particles passed through the gold foil undeviated.
2. It is an atom of lead. It has 82 positive protons in its nucleus and $(207 - 82) = 125$ neutral neutrons in its nucleus. It normally has 82 negative electrons orbiting the nucleus.

Page 85: Radioactive Decay

Quick Test 40

1. $^{228}_{89}\text{Ac} \rightarrow ^{228}_{90}\text{Th} + ^{0}_{-1}\text{e}$ and then

   $^{228}_{90}\text{Th} \rightarrow ^{224}_{88}\text{Ra} + ^{4}_{2}\alpha$

2. Alpha.
3. $A = N/t = 3600/60 = 600$ Bq.

Page 87: Fission, Fusion and $E = mc^2$

Quick Test 41

1. Both processes result in a loss in mass and the production of energy using $E = mc^2$. Spontaneous fission happens without being caused by any external influence. Induced fission is caused by bombarding a nucleus with a neutron.

2. Both processes result in a loss in mass and the production of energy according to $E = mc^2$. Fission is when a large nucleus breaks down into smaller nuclei (+ neutrons). Fusion is when small nuclei join together to form a larger nucleus.
3. 1 g = 0.001 kg. $E = mc^2$
   $= 0.001 \times (3.0 \times 10^8)^2 = 9.0 \times 10^{13}$ J
4. (a) induced nuclear fission.
   (b) mass before $= (390.219 + 1.675) \times 10^{-27}$
   $= 391.894 \times 10^{-27}$ kg
   mass after $= (227.292 + 157.562 + [4 \times 1.675]) \times 10^{-27} = 391.554 \times 10^{-27}$ kg.
   Loss in mass $= 3.4 \times 10^{-28}$ (kg)
   energy released $= E = mc^2 = 3.4 \times 10^{-28} \times (3.0 \times 10^8)^2 = 3.06 \times 10^{-11}$ J

Page 89: Dosimetry and Safety

Quick Test 42

1. $H = DQ = 3.6 \times 20 = 72$ μSv.
2. $80 \rightarrow 40 \rightarrow 20 \rightarrow 10$ means three halvings. Thickness required $= 3 \times 15 = 45$ mm.

Page 93: Uncertainties

Quick Test 43

1.

Quantity	Unit
mass	kilogram (kg)
weight	newton (N)
electric charge	coulomb (C)
capacitance	farad (F)
period	second (s)
absorbed dose	gray (Gy)

2. 56×10^{-12} A OR 5.6×10^{-11} A.
3. 480×10^{-9} m OR 4.8×10^{-7} m
4. (a) 3.4×10^2 ms^{-1}
   (b) 2.5×10^3 kV OR 2.5×10^6 V
   (c) 4.7×10^3 μF OR 4.7×10^{-3} F
5. This uncertainty is due to inaccuracy in the apparatus. Any readings taken from the scale are likely to be greater than the correct lengths because the end of the metre stick is worn away. Since all the readings tend to be greater than the true value, this is a systematic uncertainty.

Index